WITHDRAWN

Accession no.

D1323446

Modality

OXFORD SURVEYS IN SEMANTICS AND PRAGMATICS

GENERAL EDITORS: Chris Barker, New York University, and Chris Kennedy, University of Chicago

ADVISORY EDITORS: Kent Bach, San Francisco State University; Jack Hoeksema, University of Groningen; Laurence R. Horn, Yale University; William Ladusaw, University of Southern California; Beth Levin, Stanford University; Richard Larson, Stony Brook University; Anna Szabolsci, New York University; Mark Steedman, University of Edinburgh; Gregory Ward, Northwestern University

PUBLISHED

1 *Modality*
Paul Portner

IN PREPARATION

Reference
Barbara Abbott

Intonation and Meaning
Daniel Büring

Questions
Veneeta Dayal

Indefiniteness
Donka Farkas

Aspect
Hana Filip

Lexical Pragmatics
Laurence R. Horn

Mood
Paul Portner

Dimensions of Meaning
Chris Potts

Modality

PAUL PORTNER

LIS LIBRARY

Date	Fund
17	7	09	19

Order No
2045928

University of Chester

OXFORD

UNIVERSITY PRESS

OXFORD
UNIVERSITY PRESS

Great Clarendon Street, Oxford OX2 6DP

Oxford University Press is a department of the University of Oxford.
It furthers the University's objective of excellence in research, scholarship,
and education by publishing worldwide in

Oxford New York

Auckland Cape Town Dar es Salaam Hong Kong Karachi
Kuala Lumpur Madrid Melbourne Mexico City Nairobi
New Delhi Shanghai Taipei Toronto

With offices in

Argentina Austria Brazil Chile Czech Republic France Greece
Guatemala Hungary Italy Japan Poland Portugal Singapore
South Korea Switzerland Thailand Turkey Ukraine Vietnam

Oxford is a registered trade mark of Oxford University Press
in the UK and in certain other countries

Published in the United States
by Oxford University Press Inc., New York

© Paul Howard Portner 2009

The moral rights of the authors have been asserted
Database right Oxford University Press (maker)

First published 2009

All rights reserved. No part of this publication may be reproduced,
stored in a retrieval system, or transmitted, in any form or by any means,
without the prior permission in writing of Oxford University Press,
or as expressly permitted by law, or under terms agreed with the appropriate
reprographics rights organization. Enquiries concerning reproduction
outside the scope of the above should be sent to the Rights Department,
Oxford University Press, at the address above

You must not circulate this book in any other binding or cover
and you must impose the same condition on any acquirer

British Library Cataloguing in Publication Data
Data available

Library of Congress Cataloging in Publication Data
Portner, Paul.
Modality / Paul Portner.
p. cm. – (Oxford surveys in semantics and pragmatics)
ISBN 978–0–19–929243–1 – ISBN 978–0–19–929242–4
1. Modality (Linguistics)   2. Modality (Logic)   I. Title.
P299.M6P67 2009
415–dc22   2008041664

Typeset by SPI Publisher Services, Pondicherry, India
Printed in Great Britain
on acid-free paper by
CPI Antony Rowe, Chippenham, Wiltshire

ISBN 978–0–19–929242–4 (Hbk.)
ISBN 978–0–19–929243–1 (Pbk.)

1 3 5 7 9 10 8 6 4 2

Contents

List of Figures and Tables

Figures

Tables

Acknowledgments

Writing this book has been one of the most enjoyable scholarly exercises I've ever undertaken. It has given me the opportunity to study a wide range of literature, to learn a tremendous amount, and finally to share my opinions with an audience. So I would like to express my appreciation to all of those who have helped this project to happen.

To begin with I thank the editors of the series *Oxford Studies in Semantics and Pragmatics*, Chris Kennedy and Chris Barker, for the confidence they have shown in me and their support in defining and then redefining the project. John Davey at Oxford University Press has combined a knowledge of the publishing world with an understanding of the vagaries of research and writing. I thank Georgetown University for essential support in the form of a year of sabbatical and research leave, and I very much appreciate the welcome I received from the National Tsing Hua University during that time. The beautiful Tsing Hua campus was the perfect place to concentrate while the largest part of this book was written. I also thank the Stanford Linguistics Department for allowing me to be a visiting scholar during several summers, thereby giving me access to the Stanford Libraries.

I have made good use of my access to the knowledge and wisdom of fellow linguists while working on this book. I particularly thank the following people for more extended discussions: Raffaella Zanuttini, Miok Pak, Elena Herburger, Graham Katz, Steve Kuhn, Valentine Hacquard, and Dylan Tsai; I especially thank Steve Kuhn and Valentine Hacquard for extensive comments on the pre-final manuscript. Several research assistants and students provided important logistical help, including Goldie Ann Dooley, Eric Flaten, and Hui Chin Tsai. I'm probably forgetting one or more important conversation I had along the way, and for that I apologize. I presented small parts of this work at the University of Texas at Austin, the International Conference on Linguistics in Korea at Seoul National University, the Workshop on Philosophy and Linguistics at the University of Michigan, National Tsing Hua University, the University of Maryland, and Rutgers University. I thank all the inviters, organizers, and participants.

My family has been of tremendous help in allowing me to put in the time and effort to finish this work. My wife Sylvia, my parents Mary and Allen, my in-laws Klaus and Linna (over the last year, especially Linna) have given me the peace of mind that the two little boys Noah and Ben are happy and nourished in all ways while I work.

Paul Portner
Georgetown University
May 2008

General Preface

Oxford Surveys in Semantics and Pragmatics aims to convey to the reader the life and spirit of the study of meaning in natural language. Its volumes provide distillations of the central empirical questions driving research in contemporary semantics and pragmatics, and distinguish the most important lines of inquiry into these questions. Each volume offers the reader an overview of the topic at hand, a critical survey of the major approaches to it, and an assessment of what consensus (if any) exists. By putting empirical puzzles and theoretical debates into a comprehensible perspective, each author seeks to provide orientation and direction to the topic, thereby providing the context for a deeper understanding of both the complexity of the phenomena and the crucial features of the semantic and pragmatic theories designed to explain them. The books in the series offer researchers in linguistics and related areas—including syntax, cognitive science, computer science, and philosophy—both a valuable resource for instruction and reference and a state-of-the-art perspective on contemporary semantic and pragmatic theory from the experts shaping the field.

Modality is a fundamental topic in the study of meaning, as it underlies one of the most significant features of human language: the capacity to convey information about objects and events that are displaced not only in time and space but also in actuality or potentiality. It is also one of the most challenging topics for an introductory treatment, since the empirical details are both subtle and endlessly intricate, and the formal approaches are correspondingly complex. We are fortunate to have Paul Portner's depth of experience with this topic, breadth of scholarship, and clarity of mind informing the inaugural volume in this series.

Chris Barker
New York University

Christopher Kennedy
University of Chicago

1

Introduction

This is a book about semantic theories of modality.

I am not too comfortable trying to define modality, but a definition provides a useful place to start: modality is the linguistic phenomenon whereby grammar allows one to say things about, or on the basis of, situations which need not be real. Let's take an example: I say "You should see a doctor." I am saying something about situations in which you see a doctor; in particular, I am saying that some of them are better than comparable situations in which you don't see a doctor. Notice that what I say can be useful and true even though you do not see a doctor. Thus, what I say concerns situations which need not be real.[1]

This definition does not make plain exactly which features of language are associated with modality. For example, is the past real? That's a hard question, and if the past is not real, according to the definition the past tense counts as a modal expression. Another example: what is it to be kind? I help those around me with a feeling of happiness in my heart. But my life is easy, and acting this way does not lead to any difficulties in my own life. Is this enough to make me kind? Or does kindness also require that I would continue to help, with that feeling of happiness in my heart, even if doing so were to lead to difficulties for myself? If the latter, the word *kind* involves modality. It seems that modality is not something that one simply observes, but rather something that one discovers, perhaps only after careful work.

As a practical matter, the right way to discover modality is to begin with some of the features of language which most obviously involve modality, to understand these as well as possible, and then to see whether that understanding is also fruitful when applied to new features of language. In semantics, this strategy has proceeded by first

---

[1] Some, like David Lewis (1986b), would say that the situations in which you see a doctor are real, but not actual. I don't mean to make a distinction between what's real and what's actual here. My goal is just to convey what modality is as clearly as possible.

studying certain auxiliary verbs like *must*, certain adverbs like *maybe*, and certain adjectives like *possible*, since the meanings of these obviously have to do with situations that are not real. Semanticists then develop theories of these words and the constructions they occur in, and finally they see whether these theories are useful in understanding the meanings of other words, phrases, and constructions. To the extent that the theories in question—let's call them semantic theories of modality—also contribute to a better understanding of new phenomena, we learn that those phenomena involve modality as well.

After decades of research, linguists have identified very many modal words, phrases, and constructions. (Of course, as with any scientific enterprise, almost any conclusion is subject to revision. We might have made a mistake in calling something modal, and all we can do at any point in time is to apply our best judgment.) In the remainder of this chapter, I am going to do some initial work listing and classifying the various expressions of modality.

An important traditional way of classifying varieties of modality is into the categories of EPISTEMIC and DEONTIC. Epistemic modality has to do with knowledge, as in (1a), while deontic modality has to do with right and wrong according to some system of rules, as in (1b):

(1)   (a)   John must be sick.
      (b)   John must apologize.

Notice that the same word, here *must*, can be epistemic in one case and deontic in another. It turns out that the division of modals into epistemic and deontic categories is much too simple. For example, (2) is about neither knowledge nor rules:

(2)   If you like chocolate, you simply must try this ice cream.

We'll see various improved classifications based on the epistemic/deontic distinction in Chapters 3 and 4.

Another way of dividing up the modal forms is into the following three categories:

1. SENTENTIAL MODALITY is the expression of modal meaning at the level of the whole sentence. This includes the traditional "core" modal expressions: modal auxiliaries and sentential adverbs like *maybe*. Syntactic theories place elements expressing sentential modality above the level of the predicate, for example in IP or CP (or the corresponding structures in other theories). Some readers

may find it useful to compare with the more common term "sentential negation," since I am aiming for a similar concept.

2. SUB-SENTENTIAL MODALITY is the expression of modal meaning within constituents smaller than a full clause, for example within the predicate (e.g., by verbs) or modifying a noun phrase (e.g., by adjectives). There is some slippage between this category and sentential modality. Although the adjective *possible* is technically a representative of sub-sentential modality, the structure *It is possible that S* is often discussed alongside expressions of sentential modality. The reason for this is that semanticists often assume that the semantic relationship between *possible* and its complement clause is identical to that between a modal auxiliary such as *may* and the *may*'s own clause. That is, it is assumed that both can be schematized as *MODAL+S*, and that it's not important whether there is a clause boundary between the modal and S.

I also include verbal mood (in particular, the indicative and subjunctive) within sub-sentential modality, since contemporary theories view mood as largely determined by sub-sentential modality higher up in the structure. For example, in many languages a sentence of the form *It is possible that S* will require that S has a subjunctive form, and the reason that the subjunctive is needed here has to do with the kind of modality expressed by *possible*.

3. DISCOURSE MODALITY is any contribution of modality to meaning in discourse which cannot be accounted for in terms of a traditional semantic framework. Since the vast majority of work on the semantics of modality builds on a truth-conditional approach to sentence meaning, we may tentatively say that discourse modality is any modal meaning which is not part of sentential truth conditions. One of the major themes of Chapter 4 is that many sentential modal elements also involve discourse modality. There are also important connections between sub-sentential modality and discourse modality; for example, the choice of verbal mood in root sentences (typically indicative, but sometimes the subjunctive or some other mood) is dependent on the role which that sentence has in the discourse.

This book will focus on theories of sentential modality. I do not privilege sentential modality because I think it is somehow more important or more central to our understanding of modality generally. In fact, this book was originally planned to be about all three types. But it turned out to be impossible to discuss all three with enough detail to be useful, and so the project was split in half. This book is primarily about sentential modality, and a subsequent one, *Mood* (Portner, forthcoming), will examine theories of sub-sentential and discourse modality. Though the present book is primarily about sentential modality, it will examine sub-sentential and discourse modality wherever necessary. In particular, Chapters 4–5 have a lot to say about sub-sentential and, especially, discourse modality. Nevertheless, a proper discussion of these topics will have to await the later work.

Given this somewhat principled and somewhat pragmatic delineation of my subject matter, let us turn to a (necessarily partial) catalogue of linguistic phenomena which appear to involve modality. Unless otherwise noted, I discuss all of the types of sentential modality mentioned below in this book, and none of the types of sub-sentential or discourse modality.

1. Sentential modality
   (a) Modal auxiliaries: *must, can, might, should,* and the like.
   (b) Modal verbs: The semi-modals of English (e.g., *need (to), ought (to)*) and verbs in other languages which do not meet the criteria for being an auxiliary in the English sense (e.g., Italian *potere* "can/may", *dovere* "must"), provided that their proper syntactic analysis places them above the level of the predicate. (If any language has modal verbs which reside in the core predicate, in the manner of an ordinary lexical verb, they will be included in the category of sub-sentential modality.)
   (c) Modal adverbs: *maybe, probably, possibly,* and so forth.
   (d) Generics, habituals, and individual-level predicates:
       i. Generics: *A dog is a wonderful animal.*
          GENERIC sentences have to do with the characteristics which are associated with the members of a group (dogs in this example). They come in several varieties, and only some of them will come into the discussion here. See Krifka et al. (1995) for a useful overview.
       ii. Habituals: *Ben drinks chocolate milk.*
          HABITUAL sentences have to do with what the characteristic behaviors of an individual. Krifka et al. (1995) provide a useful introduction to habituals. I do not discuss habituals

explicitly here, but they are very similar to the type of generic which I do discuss.

iii. Individual-level predicates: *Noah is smart.*

An INDIVIDUAL-LEVEL PREDICATE denotes a stable property, as opposed to a a property derived from a time-bounded occurrence (see Milsark 1977; Carlson 1977; a predicate which is not individual-level is STAGE-LEVEL.) Chierchia (1995b) argues that such predicates introduce genericity, and so they are modal if generics are. I don't discuss individual-level predicates here, but if Chierchia is right, the analysis of generics applies to them.

(e) Tense and aspect:

  i. The future, in particular sentences with *will*.

  ii. The use of the past to express "unreality," as in *Even if Mary stayed until tomorrow, I'd be sad.*

  iii. The progressive: *Ben is running.*

  iv. The perfect: *Ben has eaten dinner.*

(f) Conditionals (*if...*, *(then)*...sentences). Because they are such a complex topic, and because good guides already exist for the philosophical part of the literature (Edgington 1995, 2006, Bennett 2003) I will not discuss conditionals at length. I do mention their place in one important theory of modality in Chapter 3, and discuss their relation to modality more generally in Section 5.2.

(g) Covert modality: *Tim knows how to solve the problem.*

A sentence which expresses modal meaning, even though it seems that no overt material in the sentence expresses that meaning, can be said to display COVERT MODALITY. The example above, from Bhatt (1999, (1a)), exemplifies covert modality because it means "Tim knows how he *can* solve the problem." Generics and habituals can be seen as representing covert modality as well, but the term is usually applied to examples with infinitives. I do not discuss such cases here; see Bhatt's work for details.

(h) There are many linguistic constructions which have been proposed to involve modality as part of their meaning, but for which this conclusion is not generally accepted. An example at the sentential level is disjunction (e.g., Zimmermann 2000). When they build on the theory of modality, but have not contributed to it in a significant way, I won't discuss such cases here.

(i) When we look at semantic work on languages which have not received a great deal of study from linguists, it becomes apparent that we do not know all of the types of modal meaning. See, for example, Inman (1993) and H. Davis et al. (2007). (A similar point can be made about sub-sentential modality; see Tonhauser 2006, for example. Cross-linguistic work has been more central in the study of discourse modality, especially evidentiality.) I hope that the present book, by elucidating the theory of modality generally, proves helpful to students wishing to study such topics. However, the recent interesting work in this area is too diverse to cover thoroughly in a book of this kind.

2. Sub-sentential modality

(a) Modal adjectives and nouns: *possible, necessary, certain, possibility*, and so forth.

Though these do not technically involve sentential modality, as mentioned above they are usually analyzed along with sentential modals. In this book, I generally follow this standard practice, though I also make some comments on special issues which relate to these forms in Section 3.1.3.

(b) Propositional attitude verbs and adjectives: *believe, hope, know, remember, certain, pleased*, and many others.

Certain occurs in this list and the preceding one. In the form *It is certain that* . . . , we'd call it a modal adjective; in the form *John is certain that* . . . , a propositional attitude adjective. The difference is in whether the sentence mentions an "attitude holder," here John.

Since Hintikka (1961), the tools of modal logic, in particular the semantics for modal logic based on possible worlds, have been applied to a range of propositional attitude verbs.

(c) Verbal mood, in particular the indicative and subjunctive. Verbal mood is usually analyzed as being grammatically dependent on modality—typically sub-sentential or discourse modality—expressed elsewhere in the sentence or context. Verbal mood will be discussed briefly in Section 5.3.

(d) Infinitives. Infinitives were mentioned above as involving covert modality. They also sometimes behave in a way very similar to verbal mood (see Portner 1997).

(e) Dependent modals. Sometimes a sentential modal, e.g. a modal auxiliary, functions in a way similar to verbal mood, for example, *I'd be surprised if David **should** win*. Such DEPENDENT MODALS will be discussed briefly along with verbal mood.

(f) Negative polarity items. NEGATIVE POLARITY ITEMS (NPIs) are words and phrases which cannot occur freely, but rather must be licensed by another element, canonically negation, elsewhere in the sentence. For example, *David will *(not)* ***ever*** *leave*. While there are many theories of NPIs in the literature, one of them (Giannakidou 1997, 1999, 2007) argues that they are dependent on modality expressed elsewhere in the sentence in a manner closely related to verbal mood.

3. Discourse modality

(a) Evidentiality. While EVIDENTIALITY may be defined in functional terms as a speaker's assessment of her grounds for saying something, when semanticists study evidentiality they typically are interested in a narrower topic, the meanings of functional elements (for example, a closed set of affixes) which express evidential meaning. Some scholars have argued that evidentiality is a kind of sentential modality, while others treat it, in effect, as discourse modality. Evidentiality is discussed several times in this book, in particular in Sections 4.2.2 and 5.3.

(b) Clause types. Every language has sentences which are conventionally associated with the functions of asserting a proposition, asking a question, or requiring that someone perform or refrain from an action. These are the declarative, interrogative, and imperative sentences, respectively. These categories, and sometimes others as well, are known as CLAUSE TYPES or SENTENCE TYPES. I briefly discuss sentence types in Section 5.3.

(c) Performativity of sentential modals. Some sentential modals have discourse-oriented components to their meanings which cannot be derived from their basic sentential semantics. I label this property PERFORMATIVITY, and it constitutes a major theme of Chapter 4.

(d) Modality in discourse semantics. Studies of sentential and sub-sentential modals have provided important insights into the nature of discourse meaning. Most prominently, several scholars including Roberts (1987, 1989) have studied MODAL

SUBORDINATION.[2] Modal subordination is a pragmatic phenomenon in which one sentence involving (sentential) modality affects the interpretations of subsequent modal sentences.
For example: *John might go to the store. He should buy some
fruit*, where the second sentence means "If he goes to the store,
he should buy some fruit."

That's a lot to cover, even in two volumes, so I won't waste any more
words on the introduction.

---

[2] Modal subordination might not fall perfectly under the definition of discourse modality, since its effect is on truth conditions. But it is an open question whether it is an effect on
sentential truth conditions (in context), as opposed to an effect on the truth conditions of
the discourse of which the modal sentences is a part. In other words, the issue is whether it is
the result of "ordinary" context dependency, or whether it comes about essentially through
the rules which interpret discourses.

2

Modal Logic

2.1 Why modality is important to logic and to semantics

Why are we so interested in the semantics of modal words? An easy answer for a linguist to give might be that we're interested in the semantics of all kinds of words, and we might as well attend to the modal ones now rather than later. From this perspective, we could have just as well begun with words expressing family relations or describing protein sources. Surely there are perspectives on language according to which these portions of the vocabulary have as much of a claim on our attention as the modal expressions, but such perspectives do not motivate the vast majority of contemporary work in semantics. Instead, semanticists tend to think that modality is really very important. Why?

I like cute animals: pandas, pigs, praying mantises. But by studying the meanings of the words *panda*, *pig*, and *praying mantis*, it's unlikely I'll achieve even a glimmer of the pleasure I feel in learning about them, or better yet watching them. I'd better leave my interest in animals as a hobby separate from my linguistics.

I like to know things. So when I learn something new, I like to figure out what further consequences I can deduce, so that I can learn even more true things. And moreover, when you try to convince me of something, I want to evaluate the reasons you offer me for believing it. Phrases like the following are likely to come up:

(3)   So this implies that dogs are mammals.

(4)   No, that is impossible, and here's why...

My knowledge of English tells me that if I am convinced that X, and I am convinced that X *implies* Y, I can know more of the truth by believing Y too. However, sadly, I don't always know when X implies Y, as opposed to when it just seems like X implies Y. The number of mistakes I've made in life tells me that my judgment of such things is flawed. It would be a good idea to study the concept of implication, and likewise for the concepts expressed by *must*, *impossible* and similar words.

I also want to do the right thing. Lots of people tell me what is the right thing to do in various situations, and they say things like:

(5)   Since you are hungry, what you ought to do is eat some lunch.

(6)   The teachings of the one we revere tell us that you are permitted to study linguistics.

(7)   Don't complain!

Even with lots of good advice about what is the right thing to do, I sometimes get confused. Perhaps I must do X and I ought to do Y, but I don't have the time to do both! It would be a good idea to better understand how to figure out the right thing to do in the midst of complex moral principles.

LOGIC is the study of systems of reasoning. MODAL LOGIC is the area of logic which specifically focuses on reasoning involving the concepts of necessity and possibility. As I use the term, modal logic includes the logics of all modal concepts, and so includes deontic logic (the logic of obligation and permission) and epistemic logic (the logic of knowledge), among others. One can also use "modal logic" in a narrow way which leaves out deontic modality, epistemic modality, and many other linguistically interesting types; on the narrow conception, modal logic covers the concepts of necessary and possible truth (ALETHIC MODALITY).[1] Through the development of modal logic, scholars have been able to better understand these philosophically important concepts which are both difficult to employ correctly and difficult to analyze clearly and objectively.

We've seen why logicians have been interested enough in modality to develop a logic of it, and in a moment we'll see what modal logic has to contribute to the linguistic study of modality. But note that modal logic is not by any means the same thing as the linguistic analysis of modality. Modal logic is concerned with better understanding the concepts of implication, necessity, obligation, and the like, especially as they occur

---

[1] It can be hard to determine precisely which uses of modals are supposed to count as alethic, but if you put in the word *true*, as in *It is necessarily true that...*, the alethic reading tends to stand out. The motivation for the narrow conception may be an assumption that alethic modality is the most basic type of modality, in terms of which the other varieties may be defined, rather than just one type among many.

in patterns of reasoning. It's not about the meanings of the natural language expressions like *must, possible,* and *ought.* In fact, in doing logic we often try to forget about the words we normally use to express these concepts, since doing so allows us to better focus on the system of reasoning itself. In a practice akin to using F in physics to stand for a quantity related to—but ultimately different from—our everyday concept of force, in modal logic we use symbols as well. Commonly we write $\Box$ (or sometimes L) when exploring reasoning with *must, ought,* and *necessary* (among others), and we use $\Diamond$ (or M) when exploring reasoning with *might, may,* and *possible* (among others). Once begun, the study of $\Box$ and $\Diamond$ takes on a life of its own, and logicians get to know a great deal about what kinds $\Box$'s and $\Diamond$'s there can be, without any pretense that this investigation has anything to do with the semantics of human language. Like logic in general, modal logic has a side which is rather similar to abstract mathematics.

Though modal logic is not the same thing as the linguistics of modal expressions, the deeper understanding which modal logic has given us of the concepts of implication, necessity, obligation, and the like, shows us why modal expressions demand the linguists' attention: These concepts both have practical importance and are very different from non-modal concepts. (The remainder of this chapter will be devoted showing why this is so.) Therefore, a semantic theory which does not attend to modality will be radically simpler than one which does, and so will provide a much less accurate overall picture of the nature of meaning in human language. In this way, modal vocabulary is very different from the cute animal vocabulary; if we leave out the cute animals, we'll miss some interesting details, but the parts of the theory which explain how we talk about cute people and ugly animals will presumably be pretty similar to what we'd need to take care of the cute animals as well. This is why modality is one of the most important topics in semantics, and cute animal vocabulary is not.

2.2 Some basic ideas from modal logic

2.2.1 Frames and models

Here is a certain kind of ant which identifies its sex by smell. Every ant gives off either a female smell or a male smell at all times. Suppose that a certain ant wants to know if it is a female or a male. All it can do is smell. But since it works in the nest and is surrounded by many other ants at all times, it can't just smell itself; the smells of all the nearby ants are mixed together. However, the ant does know that its own scent is among all of the ones it smells mixed together. Therefore it knows (8):[2]

(8)   If every ant which I smell is female, then I am female.

Not only does our ant know (8); every ant is in a position to know it, provided that he or she realizes that its own scent is among the scents that it can smell.

This ant also wants to know if it has a white spot on its back, but since it cannot see its own back, a statement similar to (8), namely (9), is not something it can be certain of.

(9)   If every ant which I see has a white spot on its back, then I have a white spot on my back.

There's a crucial difference between what the ant can figure out based on its sense of smell and what it can figure out based on its sense of sight. Because it can smell itself, it knows (8), but because it cannot see its own back, it cannot know (9). We say that, for ants, smell is REFLEXIVE: one can always smell oneself. But for ants, seeing-one's-back is not reflexive.

Examples (8)–(9) illustrate a case where an ant can know more by its sense of smell than its sense of sight, but this does not mean that smell is always better. Since ants have compound eyes, they can see in all directions at once. This means that if one ant sees another, the second one sees the first as well. We say that ant vision is SYMMETRICAL: if ant 1 sees ant 2, then ant 2 sees ant 1. Because of this, our ant knows (10). (Let's assume that the spots are so big that, if you can see a particular ant at all, you can see whether it has a spot on its back.)

(10)   If I have a white spot on my back, then every ant which I see sees an ant with a white spot on its back.

The symmetry of ant vision is what lets our ant be certain of (10). As a result, if the ant asks around and finds that not every ant it sees itself sees an ant with a white spot on its back, it can conclude that it does not have a white spot on its back.

---

[2] This way of introducing the basic ideas of modal semantics was inspired by Hughes and Cresswell's (1996) modal games. The semantics for modal logic on which the games are based is derived from the work of many scholars, including Kanger (1957), Bayert (1958), Kripke (1959), Kripke (1963), Montague (1960), and Hintikka (1961), and is especially connected with Kripke's name.

Let us review what we have seen in a more abstract format. If we are considering only the sense of smell, then for any ant (11) is true and (12) is false; in terms of vision, in contrast, for any ant (11) is false and (12) is true. (The variable P can stand for any property at all that an ant might have.)

(11)   If every ant which I sense is P, then I am P.

(12)   If I am P, then every ant which I sense senses an ant which is P.

Even more abstractly, all of this has nothing really crucial to do with ants, of course. What's essential is only that the relation between ants based on smell is reflexive and that based on vision is symmetrical. So we can say that (11) is valid on a reflexive relation and (12) valid on a symmetrical relation.

Modal logic allows us to build a general, abstract theory of the connection between notions like reflexivity and symmetry and the validity of statements like (11) and (12). In order to see how this works, let us begin by defining a modal logic language **MLL**.

(13)   Definition of Modal Logic Language **MLL**:

    1.   Atomic sentences:
       An infinite number of variables are sentences of **MLL**: $p, q, r, \ldots$.

    2.   Negation:
       If α is a sentence of **MLL**, then so is $\neg \alpha$.

    3.   Conjunction, Disjunction, and Conditionals:
       If α and β are sentences of **MLL**, then so are $(\alpha \wedge \beta)$, $(\alpha \vee \beta)$, and $(\alpha \rightarrow \beta)$.

    4.   Necessity and Possibility:
       If α is a sentence of **MLL**, then so are $\Box \alpha$ and $\Diamond \alpha$.

Actually we don't need quite so many symbols in the language. The ways that logicians intend to use **MLL** means that some of them can be defined in terms of the others. For example, we can take just $\neg$, $\vee$, and $\Diamond$ as basic, and define the others as follows:

1. $\alpha \wedge \beta = \neg(\neg \alpha \vee \neg \beta)$
2. $\alpha \rightarrow \beta = (\neg \alpha \vee \beta)$
3. $\Box \beta = \neg \Diamond \neg \beta$

The last point is the only one that has specifically to do with modality. Intuitively, it embodies the assumption that something is necessary if and only if it's not possible that it's false.

Next we define two concepts which are essential for providing a semantics to **MLL**: FRAMES and MODELS.

(14)   A frame is a pair $\langle W, R \rangle$ consisting of a set W and a relation R on W.

For example:

(15)   1. W is the set of three ants {ant 1, ant 2, ant 3}.
      2. R is the "smells" relation such that:
          (a)   Ant 1 smells itself.
          (b)   Ant 1 smells ant 2.
          (c)   Ant 1 smells ant 3.
          (d)   Ant 2 smells itself.
          (e)   Ant 2 smells ant 1.
          (f)   Ant 3 smells itself.
          (g)   There are no other cases of an ant smelling another.

Above we learned that ant-smelling is reflexive, and this is formalized in terms of the notion of frame. This is an example of a reflexive frame, since for each member of W, the relation R holds between that thing and itself. Given what we said about ant-smell, every frame which aims to represent what ants can learn by smelling should be reflexive.

The members of W can be anything at all, but typically they are called "possible worlds," and so a more typical description of a frame (one which is identical to (15) as far as logical properties go) might be as follows:

(16) $F = \langle W, R \rangle$ as follows:
      1. $W = \{w_1, w_2, w_3\}$
      2. R is the smallest[3] relation on W such that:
          (a) $R(w_1, w_1)$
          (b) $R(w_1, w_2)$
          (c) $R(w_1, w_3)$
          (d) $R(w_2, w_2)$
          (e) $R(w_2, w_1)$
          (f) $R(w_3, w_3)$

---

[3] This means that for no other $x, y \in W$ is it the case that $R(x, y)$.

In what follows, I'll write $R(x, y)$ to say that x stands in the R relation to y. In logic it is customary to identify a relation with its extension, so that $R(x, y)$ could also be written as $\langle x, y \rangle \in R$. I find the $R(x, y)$ notation easier to use, especially in running text, and don't mean to imply that $R(x, y)$ is an expression in some special logical language.

We'll come back to this concept of possible world shortly; for the time being we'll stick with ants, so that we can focus on the nature of the logic without getting distracted by metaphysics. Once we understand the logic well, we will be able to see why it makes sense to construct the frames out of possible worlds.

Since $R(\text{ant } 1, \text{ant } 1)$, $R(\text{ant } 2, \text{ant } 2)$, and $R(\text{ant } 3, \text{ant } 3)$, the frame above is reflexive. A symmetrical relation (like ant vision) can give rise to a symmetrical frame, and other types of frames, based on other types of relations, will be important in what follows. Here are definitions of some types of frames:

1. **Reflexive frame:** $\langle W, R \rangle$ is a reflexive frame if (and only if)[4] for every $w \in W$, $R(w, w)$.
2. **Symmetrical frame:** $\langle W, R \rangle$ is a symmetrical frame iff for every w and $w' \in W$, if $R(w, w')$, then $R(w', w)$.
3. **Serial frame:** $\langle W, R \rangle$ is a serial frame iff for every $w \in W$, there is a $w' \in W$, such that $R(w, w')$.
4. **Transitive frame:** $\langle W, R \rangle$ is a transitive frame iff for every w, w' and $w'' \in W$, if $R(w, w')$, and $R(w', w'')$, then $R(w, w'')$.
5. **Equivalence frame:** $\langle W, R \rangle$ is an equivalence frame iff it is a reflexive, symmetrical, and transitive frame.

Now we define a model.

(17)   A model is a pair $\langle F, V \rangle$ such that F is a frame and V is a function which associates each pair of a member of W and an atomic sentence of **MLL** with the value 1 or the value 0.

The function V is called a VALUATION FUNCTION. Thus we can say that a model is a pair of a frame and a valuation function. The value 1 conventionally stands for "true," so that if $V(x, p) = 1$, this is to say that p is being treated as true for x in the model. Likewise, the value 0 conventionally stands for "false." Here's an example of a model:

(18)   1. $M = \langle F, V \rangle$ such that:

(a) F is as in (15)
(b) V is as follows:

i. $V(\text{ant } 1, p) = 1$
ii. $V(\text{ant } 1, q) = 1$
iii. $V(\text{ant } 2, p) = 1$
iv.   for all other $x \in W$ and atomic sentences ϕ, $V(x, \phi) = 0$.

---

[4] From now on, I'll use the common abbreviation "iff" for "if and only if."

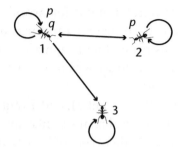

FIGURE 2.1. The model (18)

Figure 2.1 illustrates this model graphically. Ants are numbered as 1, 2, and 3 clockwise from the upper left. An arrow from one ant to another means that the first can smell the second. For example, ant 1 can smell itself and the other two ants. This has a reflexive frame (based on the reflexive relation "smells") since each ant has an arrow turning back to itself. The list of letters next to an ant represents the atomic sentences which are true for that ant. Perhaps p is "this ant is female" and q is "this ant has a white spot on its back." In that case, ant 1 is female and has a white spot on its back. Notice that, in this model, each ant can truthfully say (8), "if every ant which I can smell is female, than I am female." (Ant 1 smells two female ants, itself and ant 2, and one male ant, ant 3; because not every ant it smells is female, it can't decide whether it's female or not, and so on the basis of the definition of →, (8) comes out as true for ant 1. Ants 2 only smells female ants, and is itself female, so (8) is true for ant 2. Ant 3 smells one male ant, itself, and again (8) is true for this ant.) Of course this is as expected, because we saw earlier that this sentence should be true whenever the frame is reflexive.

Our next job is to give a formal characterization of what it is for an ant to be able to say truthfully some sentence a. We begin by starting with the simplest sentences, the atomic sentences, and then determining whether more complex sentences are true for each ant based on simpler ones. If a is an atomic sentence, the ant can say a truthfully iff V assigns 1 to the pair of that ant and a. For example, $V(\text{ant } 1, p) = 1$, so ant 1 can truthfully say "p". Next we want to allow for more complex sentences; for example, ant 1 can truthfully say "p and q", or as we write it with logical notation $(p \wedge q)$, since she can truthfully say both "p" and "q". What we want to do is take V and extend it from just the atomic sentences to all of the sentences of **MLL**. We will write $[\![\, a \,]\!]^{w,M}$ to indicate the truth value (1 or 0) of a in w on the model M.[5] In the ant model, $[\![\ p\ ]\!]^{\text{ant }1,M} = 1$ means that ant 1 in modal M can truthfully say "p".

(19) For any model $M = \langle\langle W, R\rangle, V\rangle$ and any $w \in W$, $[\![\ a\ ]\!]^{w,M} = 1$ iff one of the following conditions hold (and otherwise $[\![\ a\ ]\!]^{w,M} = 0$):

1. a is an atomic sentence and $V(w, a) = 1$
    (For atomic sentences, $[\![\ a\ ]\!]^{w,M} = V(w, a)$.)

2. a is of the form $\neg\beta$ and $[\![\ \beta\ ]\!]^{w,M} = 0$
    ($\neg\beta$ is true iff β is false.)

3. a is of the form $(\beta \wedge \gamma)$ and $[\![\ \beta\ ]\!]^{w,M} = 1$ and $[\![\ \gamma\ ]\!]^{w,M} = 1$
    ($(\beta \wedge \gamma)$ is true iff both β and γ are true.)

4. a is of the form $(\beta \vee \gamma)$ and $[\![\ \beta\ ]\!]^{w,M} = 1$ or $[\![\ \gamma\ ]\!]^{w,M} = 1$
    ($(\beta \vee \gamma)$ is true iff β or γ or both are true.)

5. a is of the form $(\beta \rightarrow \gamma)$ and $[\![\ \beta\ ]\!]^{w,M} = 0$ or $[\![\ \gamma\ ]\!]^{w,M} = 1$
    ($(\beta \rightarrow \gamma)$ is true iff either β is false or γ is true.)

6. a is of the form $\Box\beta$ and for all v such that $R(w, v)$, $[\![\ \beta\ ]\!]^{v,M} = 1$
    ($\Box\beta$ is true iff β is true in all members of W accessible from w.)

7. a is of the form $\Diamond\beta$ and for some v such that $R(w, v)$, $[\![\ \beta\ ]\!]^{v,M} = 1$
    ($\Diamond\beta$ is true iff β is true in some member of W accessible from w.)

Let me make some comments about the definitions of $\rightarrow$, $\Box$, and $\Diamond$.

We say that $(\beta \rightarrow \gamma)$ is true iff either β is false or γ is true; this is something close to the meaning of *if–then*, but it's clearly not the same as *if–then*. Don't worry about this problem; it's the best we can do within this simple logical system, and is the standard definition of $\rightarrow$. It is not how linguists typically analyze the meaning of *if–then* within their richer semantic theories. An elementary logic textbook will provide a justification for defining the meaning of $\rightarrow$ this way, but as we are only mining modal logic for the purposes of learning some things to apply to natural language semantics, it's not worthwhile going into it here. We'll discuss conditionals from different perspectives in Sections 3.1 and 5.2.

The descriptions of the definitions of $\Box$ and $\Diamond$ make use of the term "accessible from." Accessibility holds between members of W based on

---

[5] A common notation for logicians would be $w \models_M a$ or something similar. The $[\![\ \ ]\!]^{w,M}$ used here is the kind of notation used by linguists in developing theories of the semantics of natural language, and I use it here for the sake of consistency with later sections.

the relation R. That is, v is said to be accessible from w iff $R(w, v)$. In fact, R is often called an ACCESSIBILITY RELATION. In the case of our ant model, for example, ant 2 is accessible from ant 1 because ant 1 smells ant 2, that is $R(\text{ant 1, ant 2})$.

Under these definitions, $[\![\Box p]\!]^{w,M}$ means that p is true in all members of W accessible from w. In the case of our ant model, $[\![\Box p]\!]^{\text{ant 1},M}$ means "p is true for every ant that ant 1 can smell." Let's adopt the convention that the pronouns I and me refer to the ant who is saying the sentence. For example, when we are computing $[\![\phi]\!]^{\text{ant 1},M}$, these pronouns refer to ant 1. Then we can say that $\Box p$ means "p is true for every ant I can smell." Likewise $\Diamond p$ means that "p is true for some ant I can smell." The following formula expresses the meaning "if p is true for every ant I can smell, it's true for me":

(20) $\Box p \rightarrow p$

(20) is the **MLL** counterpart of (11), and it is commonly called **T**. (The reason why it is an important enough formula to warrant its own one-letter name will be explained in Section 2.2.4.) It is true for all three ants in our model, as is expected because our model has a reflexive frame.

The **MLL** counterpart of (12) is (21), called **B**:

(21) $p \rightarrow \Box \Diamond p$

This formula expresses the meaning "if p is true for me, then every ant I can smell can smell an ant for which p is true." **B** is not true for ant 1 in our "smelly" model, since p is true for ant 1, but ant 1 can smell ant 3, and ant 3 cannot smell any ant for which p is true. In other words, ant 1 is female, but not every ant that ant 1 smells can smell a female ant. It is not surprising that (21) isn't true on the "smelly" frame, since the reflexivity of smell did not guarantee the truth of (12). However, we would expect it to be true on a symmetrical "seeing" frame.

2.2.2 Validity

We can now go on to define different version of the crucial logical property of VALIDITY. The central idea of validity is that a sentence is valid iff it must be true on the basis of the language's syntax and semantics.[6] Because **MLL** is built entirely from vocabulary which is either logical in nature ($\wedge$, $\Box$, etc.) or is devoid of specific content (p's

---

[6] If we think about a natural language, where individual words have interesting meanings, we might distinguish those sentences which must be true on the basis of syntax and compositional semantics from the broader class of those which are true on the basis of

meaning is completely arbitrary), the valid sentences will all be valid due to logical properties. For example, $\neg(p \wedge \neg p)$ is valid, because p cannot be both true and not true, no matter what p itself means (and this intuitive fact is guaranteed by our definitions). What we are primarily interested in here are those sentences which are valid because of the meanings we assign to $\square$ and $\lozenge$.

First we will define validity with respect to a particular model, and then we will define it with respect to all models which share the same frame F.

(22) A sentence α is VALID IN A MODEL $M = \langle\langle W, R\rangle, V\rangle$ iff $[\![\alpha]\!]^{w,M} = 1$ for every $w \in W$.

(23) A sentence α is VALID ON A FRAME F iff, for every valuation function V, α is valid in the model $M = \langle F, V\rangle$.

(22) says that α is true for every member of W, given a particular model. That is, every ant agrees that α is true. (23) is more general; it says that α is true no matter which atomic sentences are true for each ant, just so long as we keep the ants and the relations among them the same. That is, you can go into Figure 2.1 and changes the p's and q's however you want (but don't introduce or remove ants, or change the arrows), and α will still be true for every ant. It should be clear that more sentences can be valid in a particular model than are valid on the frame of that model: It might just happen that in the model, p is true for every ant, and so p is valid on that model. But once we change the valuation function (keeping the frame the same), p is false for one or more of the ants. For this reason, p and other atomic sentences are never valid on a frame. But some complex formulas are going to be valid on a frame. For example, not only is **T** $(=(20): \square p \to p)$ valid in the model M described by Figure 2.1, it's valid in the frame F which we used to define M.

Finally, we can consider what happens if we allow the frames, i.e. the ants and the relations among them, to vary. The most general concept is K–validity:

(24) A sentence α is K–VALID iff it is valid on every frame.

---

syntax and semantics (both lexical and compositional). The first group we might call valid and the latter, TAUTOLOGIES. However, in the logic we're working with here, this distinction is not important.

We can also get various more refined notions by considering some subset of all the possible frames. Here are a number of the important versions of validity:[7]

(25) (a) A sentence a is T–VALID iff it is valid on every reflexive frame.

(b) A sentence a is B–VALID iff it is valid on every symmetrical frame.

(c) A sentence a is D–VALID iff it is valid on every serial frame.

(d) A sentence a is S4–VALID iff it is valid on every transitive frame.

(e) A sentence a is S5–VALID iff it is valid on every equivalence frame.

We already know that **T**, $(=(20): \Box p \rightarrow p)$ is T–valid and **B** $(=(21): p \rightarrow \Box \Diamond p)$ is B–valid. We'll learn more about why the various kinds of validity are important in Section 2.2.4.

2.2.3 Possible worlds

As far as logic is concerned, it does not matter what is in the set W. It could be ants, it could be numbers, it could be sentences, it could be a mix of different things. All that matters is how many things we have in W and what relations among them are established by R, and for this reason modal logic can be useful in analyzing any phenomenon which can be modeled in terms of a set of objects and relations among them. Nevertheless, modal logic was originally invented in order to develop a theory of reasoning using modal concepts, and in fact the logic we've developed does a decent job of this.

Consider the concept of T–validity. We have seen that **T** is T–valid.

$$\text{T: } \Box p \rightarrow p$$

Now consider the following sentences with *must* which have the form of **T**:

(26) (a) If it must be raining, then it is raining.

(b) If Mary must tell the truth, then she will tell the truth.

---

[7] The names for the various kinds of validity come from the links with axiomatic systems of the same names discussed in Section 2.2.4. However, the names of the axiomatic systems themselves are not based on anything easy to remember; they just come from various important works in the history of modal logic. See a good modal logic textbook like Hughes and Cresswell (1996) or Blackburn et al. (2001) for more of the historical and bibliographical details.

In (26a), *must* has an epistemic meaning, and intuitively the sentence has to be true. In contrast, in (26b), *must* has a deontic meaning, and the sentence is not necessarily true. This suggests that concept of T–validity would not be appropriate for giving a logic of modals in their deontic meanings, but it might be useful for giving a logic of epistemic modals.

Now consider the following sentence, commonly called D, which is both T–valid and D–valid:

(27)   **D:** $\Box p \rightarrow \Diamond p$

Here are some English sentences with the form of D, the first containing epistemic modals and the second deontic modals:

(28)   (a)   If it must be raining, then it might be raining.
        (b)   If Mary must tell the truth, then she may tell the truth.

Both of these are intuitively true. (In (28b), we are concerned with the reading wihere both *must* and *may* are deontic; in particular, *may* means "is permitted to." It also has a reading where *may* means "is possible that.") The fact that (28a) has the form of a T–valid sentence further supports the use of the concept of T–validity to create a logic for epistemic modals; the fact that (28b) is D–valid suggests that D–validity might be useful in constructing a logic of deontic modality.

Modal logic captures many of the logical properties of modal expressions. Of course in a way this is not surprising, since this is what it was designed to do, but why are these abstract concepts defined in terms of frames and models part of the picture? For example, why does the notion of T–validity seem to capture some important aspects of the meaning epistemic modals? The reason is that we can understand modal expressions to invoke particular frames which have the general properties of reflexivity, seriality, etc. These frames will have a set *W* and an accessibility relation *R* just like any frames, but the nature of *W* and *R* will be specified in greater detail. In particular *W* will be a set of POSSIBLE WORLDS and *R* will be defined in terms of linguistically relevant concepts like knowledge or rules.

The notion of possible world goes back to the work of Leibniz and plays an important role in modern logic and semantics. A possible world is a complete way that the universe could be throughout its history. For example, our universe (the "actual world") is a possible world. There are other possible worlds which are like our world except that some minor detail is changed; perhaps it's .0001 degree colder in

London today. Other worlds involve major changes. Perhaps dinosaurs never went extinct; there are lots of different possible worlds realizing this general scenario. Still others never had much to do with our world at all. For example, there are possible worlds in which only two elementary particles exist, and time is circular, with those two particles orbiting each other once for each cycle of time. Perhaps some possible worlds are not even conceivable by humans, but happily if there are such worlds, they can safely be ignored by linguists.

Given the notion of possible worlds, consider the following type of frame. In this frame R is defined in terms of the someone's knowledge. We have $R(w, w')$ iff everything that some individual i knows in w is true in w'. Let's call R an EPISTEMIC ACCESSIBILITY RELATION and F an EPISTEMIC FRAME:

(29)   **Epistemic frame**
       $F = \langle W, R \rangle$ is an epistemic frame iff for some individual i:

   - W = the set of possible worlds conceivable by humans.
   - R = the relation which holds between two worlds w and w' iff everything which i knows in w is also true in w'.

An epistemic frame is reflexive, since it's a property of knowledge that if i knows p in w, then p is true in w. For example, if I know that it's raining right now, then it is indeed raining right now. It is often said that knowledge is properly justified true belief (but see Gettier 1963), and the fact that what's known must be true implies reflexivity. We can say that a sentence is "epistemically valid" iff it is true on all epistemic frames. Since all epistemic frames are reflexive, anything which is T–valid is also epistemically valid. This justifies using modal logic to try to better understand epistemic modality. We can also ask the converse question: are all epistemically valid sentences also T–valid, i.e. are the epistemically valid sentences and the T–valid sentences identical? My personal answer is: I'm not sure.

We can define another important class of accessiblity relations, the DEONTIC ACCESSIBILITY RELATIONS. A deontic accessibility relation is one defined in terms of a particular system of rules. We have $R(w, w')$ iff all of the rules of w are followed in w'. For example, in our world the rule "No murder!" holds, and so we have R(our world, w') only if there is no murder in w'. More generally, $R(w, w')$ means that w' is a perfect world from the perspective of the rules of w.

Based on the concept of a deontic accessibility relation, we can define a DEONTIC FRAME:

(30)   **Deontic frame**

   $F = \langle W, R \rangle$ is a deontic frame iff for some system of rules r:

   - W = the set of possible worlds conceivable by humans.
   - R = the relation which holds between two worlds w and w' iff all of the rules which r establishes in w are followed in w'.

Note that in this definition, r is not just a set of rules. It is called a "system of rules," and it associates each world in W with a set of rules. Different worlds may be associated with different rules. For example, let r represent the moral precepts which humans should live by. Suppose we believe that a benevolent God is responsible for the nature of morality. God created all of the possible worlds, and associated a moral order with each one. The moral orders may differ somewhat from world to world, since the creatures living in each world may have different natures and so need different rules.

A sentence is deontically valid iff it is true on every deontic frame. A deontic frame is not necessarily reflexive, as we can see from the fact that, though "No murder!" is a moral precept of our world, there is nevertheless murder. However, we might require that every deontic frame is serial. By invoking seriality, we would be saying that it's no good to have an inconsistent set of rules, a set all of which cannot be satisfied together. Of course in reality, we may find ourselves in a situation with contradictory requirements (a "Catch-22"). According to the idea that deontic frames are always serial, these requirements do not constitute a proper system of rules.

Let us assume that a deontic frame is defined to be serial. Given seriality, we know that any sentence which is D–valid is also deontically valid; this justifies using modal logic to better understand deontic modality. Again, it's not clear whether every deontically valid sentence is D–valid, that is whether D–validity is precisely the same concept as deontic validity.

If we interpret $\Box p$ on an epistemic frame, it will give a reasonably good analysis of the sentence "In light of i's knowledge, it must be that p" or "i knows that p." Likewise, if we interpret $\Box p$ on a deontic frame, it goes a fair way towards providing an analysis of "In light of r, it must be that p" or "In light of r, p is obligatory." This success suggests that our understanding of modal sentences actually involves an understanding of alternative possible worlds. When you come to think of it, it doesn't seem so surprising that we would have some way of talking about alternative possible worlds, given the obvious fact about humans that we can imagine the world being different from how it is. In particular, the kinds of things we talk about with modal sentences seem intuitively to involve the alternative ways the world could be. For example, the whole point of specifying a law is that sometimes people don't behave in accordance with it. We know how we want things to be, even though we're quite aware that things are not that way. So the law specifies a set of possible worlds.

Let us look at these points with an example. The following law describes those worlds in which all scooter drivers wear helmets from the time the law comes into force:

(31)   Henceforth, all scooter drivers will wear helmets.

Based on this law, we now want to describe our world as one in which this law holds. Clearly this doesn't amount to our world being one of those described by (31); that is, the mere fact that we have this law doesn't ensure that all scooter drivers will in fact wear helmets. Rather, once we've introduced this law, our world in one in which an alternative world can only be "how things legally should be" if it has no helmetless scooter drivers. In other words, our world is one in which the following is true:

(32) $\Box$(*all scooter drivers wear helmets.*)
       $\approx$ All scooter drivers must wear helmets.

Thus, possible worlds provide an intuitively appealing tool for bringing out the details of what (32) means, and moreover the appeal to possible worlds lets us understand the success of the $\Box$ in modeling the meaning of deontic *must*. For these reasons, possible worlds and modal logic have been seen as an extremely appealing basis on which to develop semantic theories of modal expressions in natural language.

2.2.4 Axiomatic systems

In Section 2.2, I presented some of the ideas from modal logic which are especially important to linguistic semantics. However, there's much more to modal logic than this, and in what follows I briefly discuss one other important aspect of modal logic. This section is designed to give readers a flavor for how the ideas from modal logic on which linguists draw fit into the broader field.

The original conception of modal logic was syntactic (e.g., Lewis 1918; Gödel 1933; see Blackburn et al. 2001 for a brief historical overview). Given a modal language like **MLL**, a system of axioms and system for proving theorems from these axioms can be added, with the logic being, in effect, the set of theorems that can then be proven.[8] For example, let's consider the following logic K built on **MLL**:

(33)   Axioms of K

   (a)   If α is a valid sentence of non-modal propositional logic, then α is an axiom of K.
   (b)   **K**: $\Box(p \rightarrow q) \rightarrow (\Box p \rightarrow \Box q)$ is an axiom of K.

(34)   Rules of Proof

   (a)   Uniform Substitution
         If α is a theorem of K, then so is the result of replacing any of the variables, in a uniform way, with sentences of **MLL**.
   (b)   Modus Ponens
         If both α and $(\alpha \rightarrow \beta)$ are theorems of K, then so is β.
   (c)   Necessitation
         If α is a theorem of K, then so is $\Box\alpha$.

The first set of axioms in our modal logic, those given by (33a), is not specifically modal in nature. This just says that if something is valid in the language we had before adding modal operators, we want it to be a theorem of the modal language. This makes sense. For example, if we accept that "p or not p" ($p \vee \neg p$) is obviously true (as most people do), we are entitled to assume it as we go about proving things in modal logic.

Axiom (33b) is a specifically modal axiom. This axiom is traditionally known as **K** (note: boldface) because what distinguishes the logic K (no boldface) from other modal logics is that it has only this as a specifically modal axiom. In other words, since we are going to compare K to other modal logics which have modal axioms besides **K**, we can say that **K** defines K. Reading $\Box$ as *necessary*, this axiom says that if it's necessary that p implies q, then if it's necessary that p, it's also necessary that q. Does that sound like a principle we should accept? How about if we we replace *necessary* with *must*? K does seem pretty solid, though modal logics that do not accept it have been developed. In any case, it is worthwhile to at least develop the logic that comes about if we accept **K**, and see if it turns out to be useful.

---

[8] Hughes and Cresswell (1996) provide a straightforward introduction, on which my presentation is based.

TABLE 2.1. Some important axioms for modal logics

Axiom name	Axiom
K	$\Box(p \rightarrow q) \rightarrow (\Box p \rightarrow \Box q)$
T	$\Box p \rightarrow p$
B	$p \rightarrow \Box \Diamond p$
D	$\Box p \rightarrow \Diamond p$
4	$\Box p \rightarrow \Box \Box p$
E	$\Diamond p \rightarrow \Box \Diamond p$

The first two rules of proof are not specifically modal, and we won't dwell on them here. The third says that if you can prove some sentence a, you can immediately prove $\Box a$ as well. The idea here is that proving something is one way of finding out that it's a necessary truth. More precisely, necessitation says that if you can prove something using only the axioms and rules above, it can confidently be counted as necessary. For example, $(p \vee \neg p)$ is a theorem of K (through a trivial proof, since it is an axiom by (33a)). Since we're completely sure of $(p \vee \neg p)$, we are also completely sure that $(p \vee \neg p)$ must be true, i.e. $\Box(p \vee \neg p)$. This principle aims to capture an important connections between provability and necessity: provability implies necessity. If you're not sure that it's gotten the connection completely right, that's also fine; again, it's at least worthwhile to accept Necessitation provisionally and see where it leads us.

One can get other modal logics by using different axioms. Some of the most well-known modal logics and axioms which define them are given in Tables 2.1 and 2.2. Some of the formulas have been introduced earlier (e.g., T and B), and in a moment we'll see why.

This plethora of axioms and systems may be bewildering at first. What does each axiom mean? Should it be accepted in general, or for

TABLE 2.2. Some systems of modal logic and their axioms

System name	Axioms
K	K
T	K T
B	K T B
D	K D
S4	K T 4
S5	K T E

some specific purpose, or not at all? Which other formulas should also be considered as potential axioms? With careful thought and the kind of practice one gets through the study of modal logic, this sense of bewilderment can be greatly reduced. For example, **D** is a reasonable axiom if we are developing a deontic logic. In this context, **D** expresses the easily accepted idea that if something is obligatory, it's permissible. On the other hand, in deontic logic we would not want to accept **T**, since this would amount to saying that whatever is obligatory is true.

Despite the fact that with practice one can get a much better handle on what uses each potential axiom of modal logic has, practice alone will never give a firm answer to the question of whether a particular set of axioms defines exactly the logic we want. **K** and **D** are good axioms to choose if you want a deontic logic, but exactly how much can you prove with them? Is it just the right amount, not quite enough, or too much? So long as a given logic is analyzed purely in terms of the theorems provable within some axiomatic system, we cannot characterize the logic in any way other than in terms of those theorems. That is, we can say things like "our logic proves theorem X" or "this logic proves everything that logic proves, plus some more," but such statements don't provide an independent sense of what all of the theorems of a given logic have in common. A related weakness of the axiomatic approach is that it provides no obvious way to show that some formula is not a theorem of a given system. In contrast, with the frame semantics we can show that something is not valid by finding an appropriate model in which it is false (as, for example, we showed that **T** is not **D**-valid).

From a practical point of view, what we'd most like to know about a given logic is whether its theorems are all of the principles we'd want to accept for some particular type of reasoning. For example, is **D** the correct logic for moral reasoning? This question is impossible to answer, of course, without a clear understanding of morality itself, and this isn't something to be given by logic or linguistics. However, there are other ways of characterizing what all of the theorems of a given logic have in common which allow a much deeper understanding of the nature of the logics in Table 2.2. The most famous of these links axiomatic systems to frame semantics.

Recall the definition of K-validity in (24). This says that a sentence is K-valid if it's valid on any frame whatsoever. It is not hard to see that **K** is K-valid. **K** says that if every world is one in which $(p \rightarrow q)$ is true, then if every world is one in which p is true, every world is also one in

TABLE 2.3. Some systems of modal logic and equivalent definitions of validity in frame semantics

System name	Axioms	Frame type	Validity
K	K	all	K-valid
T	K T	reflexive	T-valid
B	K T B	symmetrical	B-valid
D	K D	serial	D-valid
S4	K T 4	transitive	S4-valid
S5	K T E	equivalence	S5-valid

which q is true. To see that this is K-valid, consider an arbitrary world w and an arbitrary frame F. Either $\Box(p \to q)$ is true in w or it's not. If it's not, then K is true (since the definition of $\to$ makes $(a \to \beta)$ true if a is false). If it is, then we ask whether $(\Box p \to \Box q)$ is true in w. Well, either $\Box p$ is true in w or it's not. If it's not, then $\Box p \to \Box q$ is true in w, so overall K is true. If it is, then we ask whether $\Box q$ is true in w as well. It is true if q is true in every world accessible from w. Let v be an arbitrary world accessible from w. We already know that $\Box p$ is true in w, and this implies that p is true in v. We also know that $\Box(p \to q)$ is true in w, and thus $p \to q$ is true in v as well. If both p and $(p \to q)$ are true in v, q is true in v as well. Since v was an arbitrarily chosen world accessible from w, we know that q is true in every world accessible from w; therefore $\Box q$ is true in w. We have therefore proven that K itself is true in w. Finally, since w was an arbitrary world, we know that K is true in any world, i.e. that K is valid on frame F. Since we didn't use any particular properties of the frame to prove K, we could have proven K to be valid in any frame whatsoever. In other words, K is K-valid.

It's easy to show that K is K-valid. What's harder to prove—but provable, as any modal logic textbook will show you—is that K is precisely the set of K-valid sentences. In other words, the set of theorems provable with K equals the set of sentences valid on any frame whatsoever. Technically, we say that K is sound and complete with respect to the set of all frames.[9] There are results like this for other modal systems as well. The nomenclature will make it easy to remember the information in Table 2.3.

[9] SOUNDNESS means that every theorem of K is K-valid. COMPLETENESS means that every K-valid sentence is a theorem of K. Together these imply that K = the set of K-valid sentences.

For example, T is sound and complete with respect to the set of reflexive frames; in other words, the set of MLL sentences which can be proven in the logic defined by the K and T axioms is equal to the set of T-valid MLL sentences. Note that there are also incomplete modal logics, logics which do not prove exactly the set of sentences which are valid on a class of frames which we can independently define; see a modal logic textbook for details (e.g. Hughes and Cresswell 1996: Ch. 9).

2.3 A linguistically realistic version of modal logic

Modal logic provides many important insights into the meanings of modal expressions, but as has been emphasized several times, its goal is not to provide a semantic analysis of natural language. Semanticists and logicians have different goals. When it comes to modality, the primary goal of the semanticist is to provide a precise theory of the meanings of modal expressions across languages which yields an accurate description of the facts and an explanation of linguistically important generalizations. The goal of the logician is to systematize and understand important features of reasoning with the concepts of necessity, obligation, and so forth. Because of this difference, modal logic ignores many important features of the meanings of modal expressions which are important to linguists. For example, it ignores the fact that in some languages epistemic modals occurs in a different position from deontic modals (see e.g. Cinque 1999). Just as importantly, modal logic does not integrate its ideas about the meanings of modal expressions into a general theory of natural language. Though sometimes the relationship between modal meanings and other meanings is discussed (for example, the relationship between modality and tense; see Section 5.1 and Thomason 2002 for a survey of work in logic), modal logic does not attempt to do this in a comprehensive way.

Though semanticists and logicians don't have the same goals, perhaps semanticists can just modify modal logic so as to make it fit with their goals. That is, can we not employ the same theoretical ideas (possible worlds, frames, etc.) with the purpose of providing a precise theory of the semantics of modal expressions across languages which yields an accurate description of the facts and an explanation of linguistically important generalizations? In this section, we'll see how far such a conception can take us. At first, we'll stay pretty close to the ideas presented in Section 2.2, and then, bit by bit, we'll add a number of refinements.

2.3.1 The Simple Modal Logic Hypothesis

We must stop working with $\square$ and $\lozenge$ and put real words into the language under analysis. We are going to be interested in sentences like the following:

(35)   (a)   Must p.
       (b)   Should p.
       (c)   May p.
       (d)   Can p.

For the time being, we'll assume that p can be any non-modalized, non-tensed English sentence, and that an appropriate syntactic analysis will place the modal in the right position. Therefore, (35a) could stand for sentences like the following:

(36)   It must be raining outside.
      = *must* (*it be raining outside.*)

(37)   Dog owners must keep their animals indoors.
      = *must* (*dog owners keep their animals indoors.*)

Among the patterns in (35), the first two contain modal operators which correspond to $\square$. They may be called NECESSITY MODALS. The second two, POSSIBILITY MODALS, correspond to $\lozenge$. However, *must* and *should* are not identical, nor are *may* and *can*. For example, both *may* and *can* can be used deontically; *may* can be used epistemically, but *can* cannot (unless it's negated):

(38)   (a)   Dogs may stay in this hotel. (deontic)
       (b)   Dogs can stay in this hotel. (deontic)

(39)   (a)   It may be raining. (epistemic)
       (b)   #It can be raining. (epistemic)
       (c)   It can't be raining. (epistemic)

Likewise, *must* can be used both both epistemically and deontically. *Should* can clearly be used deontically, and while it can be epistemic as well, this use is less natural for many speakers.

Not only are there differences between epistemic and deontic modals; there seem to be sub-varieties of these categories as well. For example, though both *may* and *can* have deontic uses, as we can see in (38) these deontic uses feel quite different. We need to develop a framework for describing and distinguishing all of the subtle and not-so-subtle differences among modal meanings.

In principle, modal logic gives us four things to work with in distinguishing the meanings of modals:

1. Whether the modal is a necessity modal (a kind of $\Box$) or a possibility modal (a kind of $\Diamond$).
2. The set W of possible worlds.
3. The accessibility relation R.
4. The valuation function V.

When semanticists apply modal logic, or something close to it, to natural language, they typically work only with points 1 and 3. The idea of point 2, that W could change, has something in common with Kratzer's theory to be discussed in Section 3.1. As for point 4, the idea that a modal could depend on V, I don't know of any proposal along these lines. Such a proposal would say that a modal can place a requirement on which atomic sentences are true, for example saying that a particular atomic sentence (e.g., *snow is white*) is true.[10]

It is worth making explicit the idea that the meanings of modal expressions in natural language can be analyzed in terms of points 1 and 3. We can call it the SIMPLE MODAL LOGIC HYPOTHESIS:

(40) **Simple Modal Logic Hypothesis:** The meaning of every modal expression in natural language can be expressed in terms of only two properties:

(a) Whether it is a necessity or a possibility modal, and
(b) Its accessibility relation, R.

The idea of this hypothesis is that the modal part of a given natural language is like **MLL** but with a number of $\Box$'s and $\Diamond$'s. We call it the *Simple* Modal Logic Hypothesis because we are leaving aside at this

---

[10] Some versions of dynamic logic have an operator ϕ?, for each formula ϕ. (ϕ? is a kind of $\Diamond$, and so we could also notate it as $\Diamond_\phi$.) The accessibility relation R for ϕ? has $R(w, v)$ iff ϕ is true at w, M and $w = v$. Note how the meaning of $\phi?\beta$ depends whether ϕ is true, and this in turn can depend on the valuation. See Blackburn et al. (2001: 13, 23) for discussion. Though this operator might be useful for the analysis of some aspects of natural language (e.g., presupposition, cf. van Eijck 1996), it doesn't seem to capture the interpretation of any modal element in natural language. For this reason I don't discuss it any further here, since the present work is about the semantics of modality (more or less as traditionally defined) in natural language, not the applications of modal logic to natural language.

point the possibility that the modal logic on which we build our analysis of natural language is something more sophisticated that **MLL**. We will discuss a theory which says that the basis should a different type of modal logic when we examine dynamic logic in Section 3.2.

2.3.2 Necessity and possibility

The following lists indicates which of the English modal auxiliaries correspond to $\Box$ and which correspond to $\Diamond$:[11]

(41) (a) Necessity modals ($\Box$): *must, should, would, (will, shall)*.
      (b) Possibility modals ($\Diamond$): *may, might, can, could*.

Sometimes people have the intuition that there should be more options than just $\Box$ and $\Diamond$. For example, deontic *should* seems to be weaker than deontic *must*, but certainly closer to *must* than to *may*:

(42) (a) You must leave right away.
      (b) You should leave right away.
      (c) You may leave right away.

However, it's not clear how to define a concept of weak necessity, as contrasting with regular strong necessity, in a way that makes sense within the basic framework of modal logic. A common intuition is that we are talking not about all accessible worlds or some accessible world, but about most accessible worlds. Suppose that $\triangle$ were such an operator.

(43) $[\![\triangle\beta]\!]^{w,M} = 1$ iff for most v such that $R(w, v)$, $[\![\beta]\!]^{v,M} = 1$

What does it mean to talk about most worlds? If the set of worlds in question were finite, then it would be relatively clear. Suppose that there are 1,000 accessible worlds (worlds v such that $R(w, v)$). Then if 900 of them made β true, we'd have a ratio of 9:1 β worlds to non-β worlds, and we'd be satisfied that $\triangle\beta$ is true; and if only 100 of them made β true, we'd have a 1:9 ratio and be sure that $\triangle\beta$ is false. Somewhere around 500–600, it might not be so clear, but perhaps we could live with the vagueness.

The problem is that it is standardly assumed that the set of worlds W we are working with is infinite. W is the set of all possible worlds— every way the world could be—or perhaps an infinite subset of this like

---

[11] It is controversial whether *will* and *shall* should be in this table. This point will be discussed in Section 5.1.

all of the possible worlds which are conceivable by humans. (To see that humans can conceive of an infinite number of possible worlds, consider first the actual world, then the actual world with one additional star in the universe, then the actual world with two additional stars, etc.) If W is infinite, then typically the set of accessible worlds will be infinite as well. Suppose that R is a deontic accessibility relation, and that v is some world accessible from the actual world. It's not relevant to any system of rules I'm aware of how many stars are in the universe, and so the world just like v except that it has one additional star is also accessible; the same goes for two stars, and so forth. Therefore the set of accessible worlds is infinite.

Since the set of accessible worlds is infinite, then typically the set of accessible worlds in which β is true will be infinite (unless it's empty). For example, suppose that β is *You leave right away*. Take some accessible world in which you leave right away. Then the worlds just like this one except that there are one, two, three, etc., additional stars in the universe are also accessible worlds in which you leave right away. Therefore, to know whether (43) is true, we're going to have to compare two infinite sets: the set of accessible worlds in which you don't leave right away, and the set of accessible worlds in which you leave right away. We are supposed to judge whether the ratio $\infty{:}\infty$ is "most." In this case, our simple minded math for "most" breaks down. One would have to seek a more sophisticated understanding "most" as it's used in (43) which takes into account infinite sets; as far as I know, this cannot be done in a way that gives the correct meaning for words like *should*.

2.3.3 Accessibility relations

According to the Simple Modal Logic Hypothesis, the only way we have at our disposal to distinguish the meanings of modals, other than classifying them according to whether they correspond to □ or ◇, is by assigning them different accessibility relations. So far, we have defined epistemic and deontic accessibility relations. At this point, we need both further kinds of accessibility relations and subtypes of the kinds we already have.

The difference in strength between *must* and *should* might be analyzed by saying that they use different subtypes of deontic accessibility relations. For example, following Bybee et al. (1994) we can distinguish sets of rules depending on how serious the consequences are for not following those rules. Given such a distinction, we might suggest that while *must* uses an accessibility relation based on a set of rules backed up by serious consequences, *should* uses an accessibility relation based on a wider set of rules, including both rules which are backed up by potentially serious consequences and those which might be violated without anything very terrible happening. On this view, the following sentences are expected to be natural, given the kinds of accessibility relations associated with *must* and *should*:

(44) (a) I'm a month late in returning the semantics students' assignments. I must grade them this weekend (potential negative impact if I don't: the semantics students are very upset).

(b) I'm also two days late in returning the syntax students' assignments. I should grade them this weekend as well (potential negative impact: the syntax students are some-what upset).

If we assume that the set of rules which form the basis of *should*'s accessibility relation includes all of those which form of the basis of *must*'s, then it follows that *must p* entails *should p*; in other words, *must p* is stronger than *should p*. This is so for the following reason. Let's call the set of worlds accessible (from world w) under *must*'s accessibility relation $R_{must}(w)$ and that accessible under *should*'s, $R_{should}(w)$. Suppose that r_{must} is the set of rules on which $R_{must}(w)$ is based and r_{should} is the set on which $R_{should}(w)$ is based. The analysis we are considering says that r_{must} is a subset of r_{should}. Notice that this implies that $R_{should}(w)$ is a subset of $R_{must}(w)$. That is, the fact that *should* cares about more rules implies that it accesses fewer worlds.

The fact that if r_{must} is a subset of r_{should}, then $R_{should}(w)$ is a subset of $R_{must}(w)$, exemplifies an important pattern, and so at this point it's important to make clear why it holds. What we want to see is that, for any sets of sets M and S, if $M \subseteq S$, then $\bigcap S \subseteq \bigcap M$. The subset relation reverses once we take intersections. Figure 2.2 illustrates a very simple example. In this figure, M contains the two sets drawn as ellipses with solid lines at the bottom, and S contains all three sets, the two indicated with solid lines plus the one drawn with a dashed line. Hence, $\bigcap S$ is the dark gray area, while $\bigcap M$ is the area which is either light or dark gray. Clearly, the dark gray area is a subset of the dark-or-light gray area.

Now we can see why *must p* entails *should p*. The reason can be seen in Figure 2.3, a Venn diagram where the ellipses indicate sets of

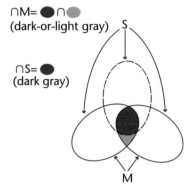

FIGURE 2.2. Intersection reverses the subset relation

possible worlds and *w* an arbitrary world which we count as the actual world. Given that the set of worlds accessed by *should* is a subset of the set accessed by *must*, if the latter is a subset of *p* (if *must p* is true), clearly the former is also a subset of *p* (then *should p* is also true). We'll examine other ways of explaining the difference between *must* and *should* in Sections 3.1.3 and 4.3.3.

The difference between *must* and *should* shows that we need at least two kinds of deontic accessibility relations. But in fact we need even more. Consider the following uses of deontic *must*:

(45)   In view of the laws of Massachusetts, drivers must yield to pedestrians.

(46)   In view of the traditions of our family, you, as the youngest child, must read the story on Christmas eve.

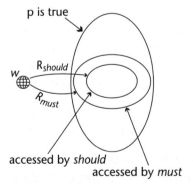

FIGURE 2.3. Deontic *must p* entails deontic *should p*

(47)   In view of the rules of student-teacher relationships, you must not yell at your teachers.

As pointed out by Kratzer (1977), we can multiply examples like these indefinitely. It seems that a given modal is compatible with many different accessibility relations, perhaps infinitely many. Therefore, when we are talking about "deontic *must*," we are not talking about something equivalent to a □ with a particular deontic accessibility relation we can specify. (One could imagine such a word existing, though it seems that no natural language has one.) Rather, we are talking about a word which is compatible with a range of accessibility relations, all of which can be classified as deontic.

Not all modality in natural language can be classified as epistemic or deontic. We need other categories. Some of the other general varieties of modality, besides epistemic and deontic, include modalities of ability, desire (called "bouletic modality"), goals (also called "teleological modality"), and perhaps history:

(48)   Fish can breathe water. (ability)

(49)   Given how you love chocolate, you should try this cake. (desire)

(50)   To get to my house, you must take a ferry. (goal)

(51)   Humanity may eventually destroy itself. (history)

We need accessibility relations corresponding to each of these types of modal. Here's a suitable definition for the desire case:

(52)   **Bouletic accessibility relation**
       R is a bouletic accessibility relation iff for some individual i, R = the relation which holds between two worlds w and w' iff all of i's desires in w are satisfied in w'.

A bouletic frame is one which has the set of possible worlds conceivable by humans as W and a bouletic accessibility relation as R. Assuming that *should* in (49) is a □ which uses a bouletic frame and that i is the addressee ("you"), the meaning of (49) is then as follows:

(53) $[\![$ You should try this cake $]\!]^{w,M} = 1$ iff the addressee tries this cake in every world in which his or her desires in w are satisfied.

For example, the addressee loves chocolate. This cake was made with good chocolate, and eating good chocolate always makes him happy. The cake is right in front of him, and nobody is going to get angry if he takes a piece. If all of this is true, it is reasonable to say, on a bouletic interpretation, that (49) is true. (53) seems to capture the intuitive reason why it is true: the bouletically accessible worlds are those in which his desires are satisfied, and all such worlds are ones in which he tries the cake.

Next let's examine the type of modality in (51). Some authors (e.g., Condoravdi 2002 and Werner 2006, see Sections 5.1.1–5.1.2) think that there is a kind of historical modality having to do with what the future might be like, given the history which has transpired so far. This is a kind of blended temporal-metaphysical modality based on the intuition that the past is fixed in a way that the future is not. For example, assuming that (51) represents historical modality, it means that everything which has happened so far does not rule out that humanity eventually destroys itself.

The accessibility relation needed to explain historical modality is developed in Kamp (1979) and Thomason (2002). It must have the properties outlined in (54):

(54)   **Historical accessibility relation**
        R is a historical accessibility relation iff for some time t, $R =$ the relation which holds between two worlds w and w' iff w and w' are identical at all times up to and including t.

The central idea here is that historical modality assumes a connection between the structure of time and the set of possible worlds, so that at any time, we can identify equivalence classes of worlds which are alike up through that time (and which only develop differences after that time). For any world, the accessible worlds are just those in its equivalence class.[12] Notice that this definition implies that $\Box p$ (on a historical interpretation) entails p, since a historical accessiblity relation is reflexive.

It's not easy to find a natural example with a clear historical interpretation. For example, (51) could be classified as epistemic rather than historical. Perhaps (55a) and (55b) are examples.

(55)   (a)   All life must someday end.
        (b)   He might have won the game.   (Condoravdi (2002, (6b)))

---

[12] The historical accessibility relation can be represented in terms of the "branching time" model of the relationship between time and modality. According to this view, time is not a line, but has a tree structure, with a fixed past (towards the root), but possibly many open futures (towards the leaves); see Section 5.1.1 for further discussion and Thomason (2002) for details.

I'm not entirely convinced by (55a); it doesn't really seem to say that, given how history has developed so far, all life will someday end. That would leave open the possibility that, if history had developed somewhat differently, we would not be in this predicament. Rather, (55a) seems to report a timeless truth. As such, it would be an example of metaphysical modality, rather than historical modality. Condoravdi's (55b) is the best candidate. This example implies that he did not win the game, and so if we assume that epistemic modals are always based on the speaker's knowledge at the speech time, the reading cannot be epistemic. (It's a delicate matter whether we should make this assumption about epistemic modals, as we'll see below, especially Section 5.1.1.) It is possible that some sentences with *will* express historical modality, but as we'll see in Sections 5.1.2, the analysis of *will* as a modal is controversial.

This discussion of varieties of deontic, bouletic, and historical modality should make it clear how we can think about and analyze varieties of modality within the Simple Modal Logic Hypothesis. Modal logic clearly provides tools which allow subtle and interesting explanations of the semantics of modal elements.

2.3.4 Problems with the Simple Modal Logic Hypothesis

Let us summarize our conclusions in this section so far. For every modal expression M:

1. Force: M is classified as either a necessity or possibility operator.
2. Accessibility Relations:
   (a) M is associated with some set A^M of accessibility relations.
   (b) These accessibility relations fall into one or more of the classes epistemic, deontic, bouletic, ability, historical, etc.
3. Meanings:
   (a) If M is a necessity operator, for each accessibility relation $R \in A^M$, a sentence of the form M(β) can be interpreted as $\Box\beta$ using the frame $\langle W, R \rangle$.
   (b) If M is a possibility operator, for each accessibility relation $R \in A^M$, a sentence of the form M(β) can be interpreted as $\Diamond\beta$ using the frame $\langle W, R \rangle$.

This group of ideas conforms to the Simple Modal Logic Hypothesis, and it does a pretty good job of capturing the meanings of sentences with modal auxiliaries. However, as a linguistic analysis, it has a very serious shortcoming: it massively over-generates. That is, it may allow us to express every meaning a given modal sentence has, but it also leads us to expect very many meanings it doesn't have. Let us see this point by examining epistemic *must*. Clearly, *must* has an epistemic interpretation; for example, suppose John is expecting Mary to stop by his house, but she's not there yet. John can say (56):

(56)   Mary must be lost.

In such a case, (56) is interpreted as a $\Box$ with the accessibility relation R_{John}:

(57) $R_{John}(w, w')$ iff everything John knows in w is true in w'.

Now suppose that Joan is also expecting Mary, and that Mary is not at Joan's house yet either. Joan also says (56). There the sentence is interpreted with the accessibility relation R_{Joan}:

(58) $R_{Joan}(w, w')$ iff everything Joan knows in w is true in w'.

So far so good. However, the theory also predicts that when John says (56), it can be interpreted with respect to R_{Joan}, and when Joan says it, it can be interpreted with respect to R_{John}. That is, John should be able to mean "Joan's knowledge implies that Mary is lost," and Joan should be able to mean "John's knowledge implies that Mary is lost." Clearly, these are not real options. When John speaks, it's about his knowledge, and when Joan speaks, it's about hers. Our theory as it stands does not allow us to rule out the incorrect interpretations.

A similar problem arises once we consider the role of time in epistemic modality. We cannot define an epistemic accessibility relation simply in terms of what some individual i knows. What we know changes over time, and so we really have to define epistemic accessibility relations in terms of what i knows at t:

(59)   **Epistemic accessibility relation**
       R is an epistemic accessibility relation iff, for some individual i and some time t, $R =$ the relation which holds between two worlds w and w' iff everything which i knows at t in w is also true in w'.

The accessibility relation $R_{John-on-Tuesday}$ which expresses what John knows on Tuesday is different from that one $R_{John-on-Friday}$ which expresses what he knows on Friday. If he says (56) on Tuesday, the meaning is based on $R_{John-on-Tuesday}$, and if he says it on Friday, the meaning is based on $R_{John-on-Friday}$. When he says it on Tuesday, the meaning cannot be based on $R_{John-on-Friday}$, and vice versa. That is, the time t has to be the time at which the sentence is spoken. This fact is not captured by the Simple Modal Logic Hypothesis as summarized in 1–3 above.

Analogous problems can be constructed for modals other than epistemic ones. Historical modality is a very clear case in point. The way (54) is set up, we have a different historical accessibility relation for each time t. But when we use a historical modal (assuming there are any), we are not free to interpret it with any historical accessibility relation we want. Rather, the time in question must be the time at which the sentence is used, at least with simple examples like (51). It seems that some innovations in our semantic system are called for.

The general issue that emerges is that it is not enough to classify each modal as epistemic, deontic, etc. In determining which accessibility relation to use, we also need to make reference to such concepts as the identity of the speaker and the time at which the sentence is used. In what follows, we will see how to build such information into the semantics of modals in a way that the unattested interpretations of (56) can be ruled out. It should be noted that this enriched system can no longer be identified with the simple modal logic presented in Section 2.2. Nevertheless, it is still close enough that most semanticists would be willing to describe it as embodying an attempt to analyze English modal auxiliaries within modal logic.

2.3.5 The indexicality of modals

We have seen at least two things which can be relevant to choosing a correct accessibility relation for a modal, other than the classification of the relation as epistemic, deontic, etc. These are the speaker and the time at which the modal sentence is used. A striking thing about these two is that they are both INDEXICAL concepts. (Some linguists would call them DEICTIC concepts.) Indexicals are elements in natural language whose meanings make essential reference to the situation in which they are used, the CONTEXT OF UTTERANCE. Simple indexical expressions include the words *I*, *now*, and *here*. The meaning of *I*, for example, is that it introduces reference to the speaker into the process of semantic composition. In the case of (60), the subject *I* refers to Joan, and so the sentence is true iff Joan is happy.

(60)   Joan: I am happy.

Following Kaplan (1989), we can distinguish two "levels" of the meaning of an expression: its CHARACTER and its CONTENT. The character of an expression is its context-independent meaning: the character of *I* is to pick out the speaker. The content of an expression is the contribution it makes to semantics within a particular context. The content of I in (60) is Joan. Character can be modeled formally as a function, at least in many cases. For example, we can think of the character of I as in (61a); then, if we call the context of Joan's utterance in (60) u_j, we calculate the content as in (61b):

(61)   (a)   The character of I = the function f_I from contexts to individuals such that, for any context c, $f_I(c)$ = the speaker in c.
       (b) $f_I(u_j)$ = Joan

When we have a sentence containing an epistemic modal, we want to make sure that the individual i and the time t referred to in the definition of the epistemic accessibility relation (59) are the speaker and the time at which the modal is used. That is, we want $f_I(c)$ and $f_{now}(c)$ to be used in the semantic rule used to interpret the modal. One way to accomplish this formally is in terms of an accessibility relation function:

(62)   **Accessibility relation function**
       A is an accessibility relation function iff

       1.   Its domain is a set of actual and/or hypothetical contexts of utterance, and
       2.   Its range is a set of accessibility relations.

Just as we distinguish subtypes of accessibility relations, we can distinguish subtypes of accessibility relation functions. For example:

(63)   **Epistemic accessibility relation function**
       A is an epistemic accessibility relation function iff

       1. A is an accessibility relation function, and
       2.   For every context c in the domain of A, $A(c)$ = the relation which holds between two worlds w and w' iff everything which $f_I(c)$ knows at $f_{now}(c)$ in w is also true in w'.

A slightly different way of incorporating indexicality into the semantics of modals is by making the accessibility relations themselves into relations between various aspects of the context and possible worlds. For example, we could build our semantics on the relation (64):

(64) A is an epistemic accessibility relation iff for any worlds w and w', any individual i and any time t, A = the relation which holds between $\langle w, i, t \rangle$ and w' iff everything i knows in w at t is true in w'.

TABLE 2.4. Revisions to the semantic system

	Old: $[\![\ ]\!]^{w,\langle\langle W,R\rangle,V\rangle}$	**New:** $[\![\ ]\!]^{w,c,\langle\langle W,A\rangle,V\rangle}$
Frame	A set of worlds W and an accessibility relation R	A set of worlds W and an accessibility relation function A
Model	A frame and a valuation	A frame and a valuation
Parameters	A world and a model	A world, a context, and a model

In this formulation, the accessibility relation is not merely a relation between worlds. It is a relation between triplets of a world, individual and time, on the one hand, and worlds, on the other.

In order to account for the indexicality of modals, we also must revise the semantic rule used to interpret modal sentences. Previously, our interpretation function $[\![\ ]\!]^{w,M}$ had two parameters: a world and a model. A model consisted of a frame (a pair of a set of worlds W and an accessibility relation R) and a valuation function V. What we want to do is to change this a bit so that features of the context of utterance are available to the modal. Here's one way to do this: A frame is now a pair $\langle W, A\rangle$ of a set of worlds and an accessibility relation function. A model is a frame and a valuation. And the interpretation function has three parameters: a world, a context and a model. All of these changes are summarized in Table 2.4.[13] (Let me warn you that the parameters on the semantic value function $[\![\ ]\!]$ will change several more times in this book.[14] Each change reflects some new ideas concerning the the semantic theory of modality. Look on the changing parameters not as an annoyance, but rather as a convenient summary of the commitments of a particular theory.)

Now we are ready for a new semantic rule for $\square$. In terms of the idea of accessibility relation functions, defined in (62), it can be given in (65). (A parallel new rule can be given for $\diamond$ as well, but I won't bother to present it explicitly.)

(65) $[\![\ \square\beta\ ]\!]^{w,c,\langle\langle W,A\rangle,V\rangle} = 1$ iff for all v such that $A(c)(w, v)$,
$[\![\ \beta\ ]\!]^{v,c,\langle\langle W,A\rangle,V\rangle} = 1$

---

[13] It is possible to define the various notions of validity in terms of the new system, and it would make a good exercise to write out the necessary rules in full detail.

[14] Sometimes we "flatten" all of the angled brackets out, so that instead of $[\![\ ]\!]^{w,c,\langle\langle W,A\rangle,V\rangle}$, we just write $[\![\ ]\!]^{w,c,W,A,V}$. And sometimes we don't write W and V at all: $[\![\ ]\!]^{w,c,A}$.

The definition in (65) gives the same basic meaning to $\Box$ as in **MLL**, but instead of using an accessibility relation directly, it calculates one by applying the accessibility relation function A to the context of utterance c. Different varieties of modals are interpreted with respect to different kinds of accessibility relation functions: An epistemic modal uses an epistemic accessibility relation function and a deontic modal uses a deontic accessibility relation function. Various subtle differences among modals (like that between *must* and *should*) will be explained in terms of the accessibility relation functions with respect to which they are interpreted.

One final improvement is necessary before we have can say that we have a version of modal logic which is as good as it's going to get for the purposes of analyzing natural language. Right now, the model only has one accessibility relation function and one context. This means that if two modals occur in the same sentence, they will use the same accessibility relation. Any two modals in the sentence of the same force would therefore have to be synonymous. This is clearly not right:

(66)   It must be raining and so we should take an umbrella.

In (66), the first modal is epistemic, and the second is deontic. However, if we compute $[\![(66)]\!]^{w,c,\langle\langle W,A\rangle,V\rangle}$, and $A(c)$ happens to be an epistemic accessibility relation, both modals will receive an epistemic interpretation, and if it happens to be a deontic accessibility relation, both modals will be interpreted as deontic. We need to make available a number of accessibility relations.

In principle, we could produce multiple accessibility relations in two ways. One would be to have multiple accessibility relation functions. On this view, a frame would have the structure $\langle W, A_1, A_2, \ldots, A_n \rangle$, where n is the number of different kinds of accessibility relation functions needed for the language in question. (Example (66) shows that n is at least 2 for English.) The other way to get multiple accessibility relations would be to have multiple contexts, so that the context used when we assign a meaning to *must* is different from the one used when we assign a meaning to *should* in (66). If the contexts are different, a single A can assign different accessibility relations to the two contexts. Say the context for *must* is c_1 and that for *should* is c_2. Then $A(c_1)$ would be epistemic and $A(c_2)$ would be deontic. Kratzer (1978) develops the second option in a somewhat different framework, but here the first option is simpler.

Following the first option, then, a frame can be defined as a tuple consisting of a set of possible worlds and some number of accessibility relation functions.[15] Each one of them is required to be of a particular kind. For example, we might say that the first has to be a epistemic, the second deontic of the sort used by *should*, the third bouletic, the fourth deontic of the sort needed by *must*, and so forth.

(67) An English frame is a tuple $\langle W, A_1, A_2, \ldots, A_n \rangle$, where:

    1. W is a set of possible worlds.

    2. A_1 is an epistemic accessibility relation function.

    3. A_2 is a deontic accessibility relation function which takes into account both rules which are backed up by serious consequences and rules which are not.

    etc.

The interpretation function for English therefore now looks like this:

$$[\![\ ]\!]^{w,c,\langle\langle W,A_1,A_2,...,A_n\rangle,V\rangle}$$

Each modal expressions is interpreted by a rule which makes reference to a particular A. For example:

(68) (a) *Must$_E$* β is interpreted as $\Box_1\beta$, defined as follows:

     (b) $[\![\Box_1 \beta]\!]^{w,c,\langle\langle W,A_1,A_2,...,A_n\rangle,V\rangle} = 1$ iff

$$\text{for all } v \text{ such that } A_1(c)(w, v),$$
$$[\![\beta]\!]^{v,c,\langle\langle W,A_1,A_2,...,A_n\rangle,V\rangle} = 1$$

(69) (a) *Should$_D$* β is interpreted as $\Box_2\beta$, defined as follows:

     (b) $[\![\Box_2 \beta]\!]^{w,c,\langle\langle W,A_1,A_2,...,A_n\rangle,V\rangle} = 1$ iff

$$\text{for all } v \text{ such that } A_2(c)(w, v),$$
$$[\![\beta]\!]^{v,c,\langle\langle W,A_1,A_2,...,A_n\rangle,V\rangle} = 1$$

Here, epistemic *must* (= *must$_E$*) uses the first accessibility relation function, A_1 which by (67) has to be epistemic. All of the other accessibility relation functions are irrelevant to *must$_E$*. Deontic *should* uses the second, A_2, and along these lines we'd have additional rules to associate each modal expression with the correct member of the sequence $A_1, A_2, \ldots, A_n$. Table 2.5 summarizes this "Modal Logic for Linguistics."

---

[15] Another option would be to have a series of frames in the modal: $M = \langle\langle\langle W_1, A_1\rangle, \langle W_2, A_2\rangle, \ldots, \langle W_n, A_n\rangle\rangle, V\rangle$. However, unless we can find a reason why the various sets of possible worlds might differ, this would be needlessly complex.

TABLE 2.5. Summary of differences between the traditional frame-based semantics of modal logic and the revised system

	Traditional Modal Logic: $[\![\ ]\!]^{w,\langle\langle W,R\rangle,V\rangle}$	Modal Logic for Linguistics: $[\![\ ]\!]^{w,c,\langle\langle W,A_1,A_2,...,A_n\rangle,V\rangle}$
Frame	A set of worlds W and an accessibility relation R	A set of worlds W and a series of n accessibility relation functions $A_1, A_2, \ldots, A_n$
Model	A frame and a valuation	A frame and a valuation
Parameters	A world and a model	A world, a context, and a model

2.3.6 Summary

Modal logic can analyze the meanings of modal expressions in natural language by categorizing each one as a version of $\Box$ or $\Diamond$ and by assigning each an appropriate accessibility relation. In order to associate a modal expression with the right accessibility relation, we make two important changes to the system of MLL. First, we made the accessibility relation depend on the context of utterance. This allows for indexical concepts like the speaker and time of utterance to assist in finding the right accessibility relation. And second, we expand the frame to include multiple accessibility relation functions, so that modals of different kinds can each be associated with the correct one.

2.4 Looking ahead

A basic understanding of modal logic is invaluable to linguists who study modality because some of the first theoretically precise ideas about the semantics of modal expressions were developed within modal logic. Most fundamentally, the notion of possible world allows us to represent directly the intuition that modality has to do with possible but not necessarily actual situations. The modal operators are quantifiers over possible worlds; in particular, they are universal (the $\Box$) or existential (the $\Diamond$) quantifiers with a domain of quantification picked out by an accessibility relation. Within this system, logicians have discovered that formal properties of accessibility relations, such as reflexivity, seriality, and so forth, correspond to interesting logical properties in the operators themselves. For example, seriality corresponds to the **D** axiom, and therefore is appropriate for deontic logic. Such correspondences have been inspiring to linguists, since they suggest that we may

LIBRARY, UNIVERSITY OF CHESTER

be able to explain some of the semantic properties of natural language modals in a similar way.

Classical modal logic is not a semantic theory of natural language modals, but as discussed in Section 2.3 one can adapt the core ideas of modal logic for the purposes of linguistic analysis. However, few linguists believe that this adapted "Modal Logic for Linguists" provides an adequate linguistic analysis. In the next chapter, we will survey three approaches to the semantics of modality within linguistics, examining both the empirical phenomena which motivate them and the details of the theories themselves. All of these approaches diverge significantly from modal logic in how they explain the semantics of modal expressions, but some do so more than others. Kratzer's theory, discussed in Section 3.1, is the most closely related to modal logic of the three. Dynamic semantics, presented in Section 3.2, uses the concept of possible world, but does not treat epistemic modals as quantifiers. Cognitive and functional approaches, of which I give an overview in Section 3.3, don't use the notion of possible world at all.

As we look at these new semantic theories of modality, modal logic provides an important benchmark for evaluation. Are they as precise as modal logic? Do they represent a fundamental intuition about the semantics of modality as clear and compelling the key intuition of modal logic, namely, that modal statements are true or false based on alternative ways the world could be? Do they provide better empirical coverage than modal logic, or better explanations of linguistic phenomena? At the end of this discussion, we will see that some ideas drawn from modal logic remain essential to our understanding of modality in human language.

3

Major Linguistic Theories of Modality

This section will present three semantic theories of modality which have been developed within linguistics. My goal here is to provide a clear, detailed, and accurate understanding of each theory. In order to achieve this goal, we will focus on those linguistic phenomena which have been the most important in developing the theories, even though this focus necessarily means that many important issues of the semantics of modality will be left out. The next chapter will turn around this relationship between theory and research issues: my goal there will be to provide a broad understanding of the nature of modality, using theories of modality as a basis. In short, this chapter is primarily about theories of modality; the next chapter is primarily about modality itself. Of course the ultimate goal of linguistics is to unify the goals of these two chapters, to develop a theory of modality which in itself provides the best possible understanding of the nature of modality. But for the time being, while our theories are still imperfect, it is helpful to look at the topic from two angles separately.

3.1 The work of Angelika Kratzer

In a series of writings, Angelika Kratzer developed a theory of modals and conditionals which deserves to be called the "standard theory" of modality within formal semantics: Kratzer (1977, 1978, 1981, 1986, 1991a, 1991b). This work is the standard in that it is the one which a linguistically oriented semanticist is most likely to recommend to students or colleagues who wish to learn about the theory of modality, and in that it is one which can be taken on as a working assumption in semantic research without there being much risk that other scholars will object that it is a poor choice. Of course the fact that it is standard in these senses does not mean that everyone believes that it is entirely correct. As we'll see, several alternative approaches have been developed in the literature, and research in recent years has turned up a number of phenomena for which Kratzer's approach is not adequate. Nevertheless, in any attempt to understand the nature of modality in natural language, one is well served to study Kratzer's ideas in detail.

Two major ideas of Kratzer's approach are these:

1. **Relative modality**
   Modals are not ambiguous. Rather, they are relative to one or more sets of background assumptions, what she calls "conversational backgrounds." The differences among the various uses of a given modal—between epistemic and deontic *must,* for example—derive from the particular conversational backgrounds chosen in the context.

2. **Ordering semantics**
   Instead of a simple dichotomy between worlds which are accessible and worlds which are not, we work with a ranked (or partially ordered) set of worlds. This ranked set is created through the interaction of two conversational backgrounds.

In what follows, I'll discuss how each of these points represents an improvement over the "Modal Logic for Linguists" outlined in Chapter 2. We see the concept of conversational background presented most clearly in Kratzer (1977), and the ordering semantics developed in later papers, especially Kratzer (1981).

3.1.1 From modal logic to relative modality

Within any kind of possible worlds-based theory of modality, the central issue is to identify the correct set of worlds over which a particular modal expression quantifies. In Chapter 2, we accomplished this by defining a frame as consisting of a set of worlds and one or more accessibility relation functions. One sense of a modal differs from another—epistemic *must* from deontic *should,* for example—because each has its own accessibility relation function. The accessibility relation function applies to the context of use, yielding a set of accessible worlds. In this system, the set of available accessibility relation functions is set once and for all, modals are associated with these accessibility relation functions by semantic rule, and context only plays a role in determining indexical features of meaning (for example, the identity of the speaker or the time of utterance).

The essential insight of relative modality is that pragmatics plays an even more significant role than this in the association between a modal and its accessibility relation. Rather than distinguishing one sense of a modal from another by means of semantic rule, Kratzer employs fewer semantic rules and relies on pragmatics to do more work. We'll begin by seeing briefly how this idea works in a system based firmly on modal logic; in the next subsection, we'll see how Kratzer herself develops and formalizes the idea.

In the system of Section 2.3.5, pragmatic information contributes to meaning through the context parameter c. This c is conceived of as conveying the kind of information about context of utterance which is needed to interpret indexicals. Given what we've done with it, we could define a context as a sequence of indices, for example ⟨*speaker, addressee, time of utterance, place of utterance*⟩. But surely context plays a greater role in the determination of sentence meaning than this. For example, if we are talking about the fact that Mary loves broccoli, (70) naturally receives a bouletic interpretation, while if we are trying to enforce the idea that children should eat everything on their plates, it naturally receives a deontic interpretation:

(70)   Mary should eat her broccoli.

According to modal logic, this difference is an ambiguity in *should*, and context helps to resolve the ambiguity only by helping the addressee determine which of the words spelled and pronounced "should" the speaker had in mind in uttering the sentence. That is, this is a lexical ambiguity, not an example of contextual determination of meaning.

Kratzer disagrees with the proposal that *should* is ambiguous. Rather, she thinks there is a single word *should*, and its meaning is partially determined by context. But since the difference between the bouletic and the deontic interpretations in (70) is in the set of accessible worlds, this means that context determines whether the accessible worlds are the ones which meet Mary's desires or the ones in which children behave properly at mealtime. In other words, context gives us two things: indexical information, encoded in c or a sequence like ⟨*speaker, addressee, time of utterance, place of utterance*⟩, and an accessibility relation function A. When a modal is interpreted, it draws on both the accessibility relation function—it might be epistemic, deontic, or whatever—and the indexical information in c, to assign a final meaning to the modal. The semantic rule is exactly the same as (65) above, but now we think of *A* as set by context rather than as a fixed feature of the semantic system.

According to Kratzer, different modals differ from one another in terms of which accessibility relations they are compatible with. For example, *may* is compatible with both epistemic and deontic accessibility relations. *Might* is compatible with an epistemic accessibility relation, but not a deontic one. *Should* is compatible with the not-so-serious-consequences subtype of deontic accessiblity relation, but not the serious-consequences subtype (assuming for the time being that Bybee et al. 1994 are right about the difference between *must* and *should*). Kratzer (1978) assumes that when multiple modals occur in a sentence, their contexts are different, and so they can receive different meanings (the first epistemic and the second deontic, for example). This removes the need for having multiple accessibility relation functions A_1, A_2, ..., A_n among the parameters of interpretation.

3.1.2 Kratzer (1977)

The system just outlined captures the essential ideas of relative modality, but Kratzer (1977) formalizes these ideas in a way which is ultimately more appealing for linguistic theory. In this section, we'll work towards an understanding of her framework in two stages. First, we introduce the develop the key concept of "conversational background"; and then we make some further technical adjustments which lead to Kratzer's final system.

Conversational backgrounds

Though the interpretation of a modal is typically determined by context, Kratzer notes that it is possible fix the meaning by explicit linguistic means:

(71)   In view of what I know, Mary must be lost.

(72)   In view of the rules of the secret committee, Mary must leave.

The phrase *in view of what I know* determines that *must* has an epistemic interpretation, while *in view of the rules of the secret committee* gives it a deontic interpretation. How does an *in view of ...* phrase do this? Let's break it down a little bit. The phrase *what I know* seems to be a kind of relative clause, and as such it refers to the set of facts which the speaker knows. According to Kratzer, it denotes a function

f from worlds to sets of propositions: for any world w, $f(w)$ = the set of propositions which the speaker knows in w.

(73) Used in context c, *what I know* expresses that function f such that:

    (i) The domain of f is that subset of W in which the speaker of c exists;

    (ii) For any w in the domain of f, $f(w) = \{p :$ the speaker of c knows p in $w\}$

As is standard in possible worlds semantics, Kratzer takes a proposition to be a set of possible worlds. We say that p is true in w iff $w \in p$. Since $f(w)$ is a set of propositions, it is a set of sets of worlds. A common trick we'll use below is to turn this set of propositions into a set of worlds (i.e., into a single proposition) by intersecting all of the propositions in the set. Suppose that $f(w) = \{p_1, p_2, \ldots\}$. Then $\bigcap f(w) = p_1 \cap p_2 \cap \ldots$. That is, $\bigcap f(w)$ is the set of worlds in which all of the propositions in $f(w)$ are true.

Now what about *in view of*? Intuitively, *in view of* tells us that the function expressed by *what I know* is going to determine the meaning of the modal *must*. That is, there is some sort of relation between the function f in (73) and *must* which leads to the sentence having an epistemic interpretation. There are various approaches one could take towards formalizing the relation between f and *must*. For example, the relation could be mediated by syntax, with the *in view of* … phrase serving as an argument of *must* or providing an antecedent for some phonologically null element[1] which serves as its argument. These syntactically-oriented approaches would be attractive to many linguists today, but for the time being I follow Kratzer (1977) in pursuing a more semantic analysis. This approach is to use the function f as a parameter of interpretation; its relation to the modal is very similar to that holding between an accessibility relation and a modal operator within modal logic. More precisely, we say that a CONVERSATIONAL BACKGROUND f is a function from worlds to sets of propositions which serves as a parameter of

---

[1] A phonologically null element is a piece of structure which is syntactically present but which has no phonological content. Such elements can be semantically interpreted. A simple example is the subject of an imperative clause:

(i) Ø sit yourself down!

The element represented as Ø here is very much like *you* in its meaning and its grammatical properties (for example, it licenses the reflexive *yourself*), but it is not pronounced.

interpretation, and that the function of an *in view of* ... phrase is to set the conversational background for the sentence in which it occurs. If there is no *in view of* ... phrase, the conversational background is determined by context.

Given a conversational background f, it is easy to define an accessibility relation: for any worlds w and v, v is accessible from w iff every proposition in $f(w)$ is true in v. For example, suppose that f is epistemic; that is, $f(w)$ represents the set of facts known by the speaker in w. Then, v is accessible from w if every fact known by the speaker in w is true in v as well. More concisely, we can say that the set of worlds accessible from w is $\bigcap f(w)$.

Because of this close relation between conversational backgrounds and accessibility relations, it is nothing more than a technical exercise to adjust the meanings of modal operators so that they can work with the former rather than the latter. At this point, we'll follow Kratzer in making another notational simplification by suppressing the set of worlds W and the valuation function V, no longer listing them among the parameters.[2] Given these two changes, we can give the semantic rules for modals as follows:

(74) (a) If N is a necessity modal, then $[[\ N\beta\ ]]^{w,c,f} = 1$ iff for all $v \in \bigcap f(w)$, $[[\ \beta\ ]]^{v,c,f} = 1$.

     (b) If P is a possibility modal, then $[[\ P\beta\ ]]^{w,c,f} = 1$ iff for some $v \in \bigcap f(w)$, $[[\ \beta\ ]]^{v,c,f} = 1$.

In these definitions, since f is a conversational background, $\bigcap f(w)$ is the set of accessible worlds. Just as in modal logic, a necessity modal universally quantifies over accessible worlds, and the possibility modal existentially quantifies over accessible worlds. We could make the similarity to modal logic even more apparent by defining an accessibility relation from the conversational background: $R_f(w, w')$ if and only if $w' \in \bigcap f(w)$. Based on this accessibility relation, the standard modal logic meanings for $\Box\beta$ and $\Diamond\beta$ will give the expected results. All of this shows that (74) implements the essence of the traditional semantics from modal logic using conversational backgrounds in place of accessibility relations.

---

[2] We can assume that the set of worlds and valuation function are built into the brackets $[[\ ]]$ themselves. This reflects the linguist's perspective that there's just one set of words and valuation function we could be interested in: the set of worlds that are conceivable by humans and the real, correct valuation function for the language we're studying.

Finally, we can formalize the contribution of an *in view of* phrase as follows:

(75)   For any α which expresses a function from worlds to sets of propositions, $[\![\ \text{In view of } \alpha, S\ ]\!]^{w,c,f} = [\![\ S\ ]\!]^{w,c,[\![\alpha]\!]^{w,c,f}}$.

This rule says that we use the denotation of α as the conversational background to interpret S. In essence, *in view of* α shifts the accessibility relation from the one provided by context to the one expressed by α.

Entailment and compatibility

The definitions in (74)–(75) are an accurate rendition of Kratzer's (1977) theory, but in fact she formalizes her theory differently. To understand her formalization, we begin with the following definitions:

(76)   If Σ is a set of propositions and p is a proposition, then:
   (a) Σ ENTAILS p iff $\bigcap \Sigma \subseteq p$
          (i.e., iff for every world w, if $w \in \bigcap \Sigma$, then $w \in p$);
   (b) Σ is COMPATIBLE WITH p iff $\bigcap(\Sigma \cup \{p\}) \neq \emptyset$
          (i.e., iff for some world w, $w \in \bigcap \Sigma$ and $w \in p$).

(When Σ entails p, we can also say that p FOLLOWS FROM Σ.) Intuitively, we can define $\square$ and $\diamond$ in terms of these definitions: $\square p$ says that the conversational background entails p and $\diamond p$ says that the conversational background is compatible with p. For example, we have described (71) as saying that, in every world compatible with what the speaker knows, Mary is lost. This is equivalent to saying that the speaker's knowledge entails that Mary is lost. Similarly, the possibility sentence *Mary may be lost* says that Mary is lost in some world compatible with what the speaker knows; this is equivalent to saying that the speaker's knowledge is compatible with Mary's being lost. For many people, the meanings stated in terms of entailment and compatibility are somewhat easier to understand than the definitions involving quantification over worlds.

Unfortunately, we can't directly apply the definitions in (76) to our semantics for the simple reason that there is no proposition around to play the role of p. The proposition p is intuitively that expressed by the sentence β to which the modal attached. However, in the system we're working with, derived from modal logic, the semantic value of β is a truth value, 1 or 0, not a proposition. We need to recast the system so as to make sentences express propositions, sets of possible worlds, rather than truth values. This will let us take advantage of the definitions of entailment and compatibility to give a more intuitive version of the possible worlds semantics for modality. (It will also bring us in line with the mode of presentation more common in linguistic semantics.)

Thus, some more formal adjustments: the parameters of the interpretation function are a context and a conversational background: $[\![\ ]\!]^{c,f}$. The world parameter is gone, because it has been "moved" into the semantic value itself: $[\![\ S\ ]\!]^{c,f}$ is a set of worlds. Here are the new semantic rules for logical words, modified to account for this change:

(77) For any sentence a, context c, and conversational background f:
  1. If a is of the form $\neg\beta$, $[\![\ a\ ]\!]^{c,f} = \{w : w \notin [\![\ \beta\ ]\!]^{c,f}\}$
  2. If a is of the form $(\beta \wedge \gamma)$, $[\![\ a\ ]\!]^{c,f} = [\![\ \beta\ ]\!]^{c,f} \cap [\![\ \gamma\ ]\!]^{c,f}$
  3. If a is of the form $(\beta \vee \gamma)$, $[\![\ a\ ]\!]^{c,f} = [\![\ \beta\ ]\!]^{c,f} \cup [\![\ \gamma\ ]\!]^{c,f}$
  4. If a is of the form $(\beta \rightarrow \gamma)$, $[\![\ a\ ]\!]^{c,f} =$
      $\{w : w \notin [\![\ \beta\ ]\!]^{c,f}\} \cup [\![\ \gamma\ ]\!]^{c,f}$
  5. If N is a necessity modal and a is of the form $N\beta$,
      $[\![\ a\ ]\!]^{c,f} = \{w : f(w) \text{ entails } [\![\ \beta\ ]\!]^{c,f}\}$
      (i.e., $= \{w : \bigcap f(w) \subseteq [\![\ \beta\ ]\!]^{c,f}\}$)
  6. If P is a possibility modal and a is of the form $P\beta$,
      $[\![\ a\ ]\!]^{c,f} = \{w : f(w) \text{ is compatible with } [\![\ \beta\ ]\!]^{c,f}\}$
      (i.e., $= \{w : \bigcap(f(w) \cup \{[\![\ \beta\ ]\!]^{c,f}\}) \neq \emptyset\}$)

(78) A sentence a is true in a world w, with respect to a context c and a conversational background f, iff $w \in [\![\ a\ ]\!]^{c,f}$

Of course, real sentences of natural language are built out of morphemes using the rules of morphology and syntax. There is no such thing as an atomic sentence. But we can assume that what we are calling atomic sentences can be adequately analyzed in terms of a broader syntactic and semantic theory, of which the theory of modality is just one part.[3]

In this system, sentential negation is complement with respect to W, sentential conjunction is set intersection, and sentential disjunction is set union. The arrow $(\beta \rightarrow \gamma)$ is defined as $(\neg\beta \vee \gamma)$, as before. These definitions will give the same result as those in the semantics for MLL

---

[3] Unless we're talking about quite different approaches to modality like dynamic semantics or cognitive semantics, I'll assume that the broader theory is roughly as in semantics textbooks like Chierchia and McConnell-Ginet (1990) and Heim and Kratzer (1998). Portner (2005) gives an informal presentation of the formal semantics approach.

at each particular world. In other words, if before we had $[\![\alpha]\!]^{w,c,f}=1$, we now have $w \in [\![\alpha]\!]^{c,f}$.

All necessity modals follow rule (77)5 and all possibility modals follow rule (77)6. The different readings of modals are determined by the choice of conversational background. Kratzer (1977, 1981) countenances the following general categories of conversational backgrounds, but as we've seen many finer distinctions can be made:

1. Epistemic: $f(w)$ is a set of facts known in w.
2. Deontic: $f(w)$ is a set of rules in force in w.
3. Teleological: $f(w)$ is a set of goals in w.
4. Bouletic: $f(w)$ is a set of desires in w.
5. Circumstantial: $f(w)$ is a set of circumstances holding in w.
6. Stereotypical: $f(w)$ is a set of expectations concerning what w is like.

There are restrictions on the kinds of conversational background which can be used to interpret each modal. For example:[4]

(79)　(a)　$[\![\text{ must } \beta]\!]^{c,f}$ is only defined if f is an epistemic or deontic conversational background.

　　(b)　$[\![\text{ should } \beta]\!]^{c,f}$ is only defined if f is an epistemic, deontic (not-so-serious-consequences subtype), bouletic, teleological, or stereotypical conversational background.

　　(c)　$[\![\text{ can } \beta]\!]^{c,f}$ is only defined if f is a deontic, bouletic, teleological, circumstantial, or stereotypical conversational background.

The system of Kratzer (1977) does not take into account the indexicality of modals, and in this respect it suffers from the same problems as the Simple Modal Logic Hypothesis discussed in Section 2.3.4. This problem is fairly simple to fix.[5]

Properties of accessibility relations like reflexivity and seriality can be recast within the terminology of conversational backgrounds. For example, we have the following correspondences (Kaufmann et al. 2006):

---

[4] I might not have gotten the restrictions in (79) quite right, but this gives the idea. The distinctions among modals become clearer in the later theory of Kratzer (1981).

[5] The most obvious solution is to employ the same fix we used for the Simple Modal Logic Hypothesis; that is, we can replace the conversational background as a parameter of interpretation with a function from contexts to conversational backgrounds, and adjust the rules in (77) accordingly.

(80) **Reflexivity** corresponds to **Realism:** A conversational back-ground f is realistic iff for all $w \in W$, $w \in \bigcap f(w)$.

(81) **Seriality** corresponds to **Consistency:** A conversational back-ground f is consistent iff for all $w \in W$, $\bigcap f(w) \neq \emptyset$.

(82) **Transitivity** corresponds to **Positive introspection:** A conversa-tional background f displays positive introspection iff for all $w, w' \in W$, if $w' \in \bigcap f(w)$, then $\bigcap f(w') \subseteq \bigcap f(w)$.

Because of correspondences like these, we can use some of the things we learned about properties of accessibility relations in modal logic to get a better understanding of the conversational backgrounds that are appropriate for various kinds of modals. For example, since D is characterized by the class of serial frames, we expect that the conversational background for a deontic modal should be consistent. Similarly, the conversational background used for an epistemic modal should be realistic.

The combination of (77) points 5 and 6 and principles like (79) gives us the core system of Kratzer (1977). Note that Kratzer makes some changes to this theory in the latter parts of the paper, but we will not discuss them here, because they are meant to deal with problems which are taken care of in a different way in later work like Kratzer (1981), to which we now turn.

3.1.3 Ordering semantics

The theory of Kratzer (1977) shares with standard modal logic the idea that modal expressions can be categorized as either of the necessity type, giving universal quantification over accessible worlds, or of the possibility type, giving existential quantification over accessible worlds. Although most of our discussion so far has focused on how the set of accessible worlds is determined (different kinds of accessibility rela-tions, the role of context, and the role of the particular modal words used), the set itself, once given, is simple. It is just a set, and has no internal structure. The are several problems for the semantic systems of modal logic, as well as that of Kratzer (1977), which arise because of this lack of structure.

What I mean by a "lack of structure" can perhaps be better under-stood with an analogy from nominal semantics. In predicate logic and the simple extensional semantic systems of textbooks like Chierchia and McConnell-Ginet (1990) and Heim and Kratzer (1998), a noun denotes

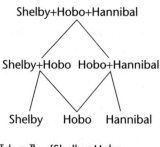

Shelby+Hobo+Hannibal

Shelby+Hobo   Hobo+Hannibal

Shelby   Hobo   Hannibal

[[dogs]] = {Shelby+Hobo,
            Hobo+Hannibal,
            Shelby+Hobo+Hannibal}

FIGURE 3.1. *Dogs* denotes the elements of a lattice

a set (or something virtually equivalent to a set like the characteristic function of a set). For example, in a particular model we might have:

(83) [[*dog*]] = {Shelby, Hobo, Hannibal}

This kind of meaning is fine for understanding a wide range of semantic phenomena, including predication, modification, and quantification, so long as we only look at singular count nouns. But it is hard to see how to apply this kind of meaning to mass nouns like *mud* and if we apply it to plural noun like *dogs*, it is hard to see how to do it in such a way that there is a difference between the singular and the plural forms. Intuitively, the problem is that mass nouns and plurals describe things which are made up of other similar or related things; in the case of mud, it is made up of smaller quantities of mud, and in the case of a group of dogs, it is made up of individual dogs. In more sophisticated theories of nominal meaning, for example, Link (1983), one can represent these part-whole relationships. For example, the plural *dogs* can be modeled as in Figure 3.1. The objects higher in the diagram, like "Shelby+Hobo," are composed of those below. Depending on one's ideas about these objects which are made out of other objects, they might be called groups, pluralities, sums, or even just sets. What's important here is that the part-whole structure illustrated by the diagram is a key part of our understanding of the meaning of *dogs*.

It is not likely that part-whole structure like that illustrated in Figure 3.1 is relevant to modals, but there is type of structure applicable to worlds which is intuitively quite relevant. Concepts like what is morally good, desirable, or probable are not all-or-nothing; they are comparative. We might understand what it is to lead a morally perfect life (or we may not), but we clearly also understand that one action may be better than another in terms of its moral status, even when the agent does not lead a completely blameless life. In terms of possible worlds semantics, we can state this by saying that, not only do we have a set of accessible worlds, but in addition we have the ability to compare worlds according to how well they fit with a moral ideal. For example, right now I can take my coffee cup and throw it into the head of a student standing outside of my office or I can sip the coffee and put the cup down on the desk. No matter which I choose, the result will not be a morally perfect life; I have done bad things in the past, and I am sure I will do more in the future. Nevertheless, it's clear that a world in which I throw the cup at the student's head is less good than one in which I do not. Not only is this clear, but it is even likely that it is more clear than the concept of a set of deontically accessible worlds which underlies the standard analysis of deontic modality in logic.

The fact that modal logic does not make use of comparison between worlds leads to a number of empirical problems for traditional modal logic and for the theory of Kratzer (1977). Most obviously, we have expressions of graded and comparative modality in language, phrases like:

(84) (a) There is a slight possibility that Mary is lost.
      (b) It is more likely that Mary doesn't want to come to your party than that she is lost.

There are also important problems with conditional sentences and with obligations which only become relevant when the situation is imperfect. In the logic literature, these are known as paradoxes of deontic logic.

Paradoxes of deontic logic

Deontic logic is an active area of research revolving to a significant extent around the need to successfully analyze arguments which are acceptable when cast in natural language but which are problematical or paradoxical when expressed in existing frameworks of deontic logic. Two well-known problems (out of many) are known as "the (Good) Samaritan Paradox" and "Chisholm's paradox" (Prior 1958; Chisholm 1963).[6] As we examine these two paradoxes, the recurring point to

---

[6] The paradoxes are closely related but slightly different. As classically presented, the Samaritan paradox has to do with the obligation to make the best of an imperfect situation, for example, by helping someone who has been injured; Chisholm's paradox concerns cases in which doing one's duty perfectly involves an obligation X, but if one fails to do one's

keep an eye on is that something goes wrong with the logic when we consider what should be done in a situation in which something bad has occurred. For example, Mary has violently robbed John. Should we tend his wounds? Roughly, the Samaritan paradox is that deontic logic (as formulated so far) answers this with "No, to tend his wounds is no better than the robbery itself." And should Mary be punished? Roughly, Chisholm's paradox is that deontic logic answers this with "She should be punished and she should not be punished."

We begin with a problem closely related to Prior's original statement of the Good Samaritan Paradox. Standard axiomatizations of deontic logic give rise to the theorem (85):

(85) $(\Box(p \rightarrow q) \rightarrow (\Box(q \rightarrow r) \rightarrow \Box(p \rightarrow r)))$

This theorem seems reasonable, but suppose we have the following values for p, q, and r:

> p: John was injured in a robbery.
> q: Someone tends the wounds John suffered in a robbery.
> r: John was robbed.

With these values, $\Box(p \rightarrow q)$ seems correct: It should be the case that, if John was injured in a robbery, someone tends his wounds. There's no way to avoid the truth of $\Box(q \rightarrow r)$, since q entails r. (Within possible worlds semantics, it's clear that in every possible world whatsoever, if q is true, r is true.) Therefore, according to the theorem, $\Box(p \rightarrow r)$ is true as well. This formula seems to say that if John was injured in a robbery, he should have been robbed, and this is a conclusion we do not want to accept.

A simpler version of the Samaritan paradox is presented by Åqvist (1967). Intuitively we would accept the following statements:

(86) (a) It ought to be that John is not injured in a robbery.
     (b) It ought to be that someone helps John, who was injured in a robbery.

---

duty perfectly, this leads to the opposite obligation $\neg X$. Because of Kratzer's influence, the Samaritan Paradox has been seen as more important by linguists, though it is not usually distinguished in a clear way from other, similar paradoxes like Chisholm's. Åqvist (1967) points out how closely related the two paradoxes are. See Hansson (1969), van Frassen (1976), and Lewis (1973) for discussion of the Samaritan paradox. The *Stanford Encyclopedia of Philosophy* entry (McNamara 2006) provides an excellent brief introduction to these and other key issues in deontic logic.

We can attempt to formalize these statements as follows:

> p: John was injured in a robbery.
> q: Someone helps John.

(87) (a) $\Box\neg p$ [formalization of (86a)]
      (b) $\Box(p \wedge q)$ [formalization of (86b)]

(87a–b) are inconsistent on the standard semantics. Obviously $(p \wedge q)$ entails p. Thus, (87b) (namely $\Box(p \wedge q)$) entails $\Box p$ (i.e., if every accessible world is a p-and-q-world, every accessible world is a p-world). Putting this together with (87a), we have $\Box p \wedge \Box\neg p$, contradicting the plausible principle that one cannot be obliged to do something and to do its opposite. (Within the possible worlds semantics for modal logic, this principle is implied by seriality of the accessibility relation for D, since seriality implies that we always have at least one accessible world, and p and $\neg p$ cannot both be true in any world.)

Though Åqvist's version of the paradox is clear and simple, it is not entirely convincing. One can doubt that (86b) is accurately represented by (87b). In particular, the non-restrictive relative clause in the former is probably not a simple conjunct. (Probably it's presupposed or a conventional implicature in the sense of Potts 2005.) Notice that we would not intuitively accept (88), a more direct analogue of (87b):

(88) It ought to be that John is injured in a robbery and someone helps John.

Thus, we should be skeptical of Åqvist's version of the logical paradox on the grounds that, since (86a) and (88) are inconsistent, it's a good result that (87a–b) are as well.

Let's look at Prior's original formulation of the Samaritan paradox next. It is stated in terms of an analysis of what it is for something to be forbidden. Intuitively, Prior (following the "Andersonian simplification" of Anderson 1956; see also Kanger 1971) treats "β is forbidden" as meaning that β is subject to sanction (e.g., punishment) of some kind. So let r in our formula be the sentence "sanction is warranted" In these terms, $\Box(\beta \rightarrow r)$ ("necessarily, if β then sanction is warranted") is a way to formalize the statement that β is forbidden. Now we can formulate the paradox:

> p: Someone tends the wounds John suffered in a robbery.
> q: John was robbed.
> r: Sanction is warranted.

With these values, we again instantiate the theorem (85): $\Box(p \to q)$ is so because p entails q. $\Box(q \to r)$ seems to be an assumption we'd want to make (it is forbidden to rob John). Therefore $\Box(p \to r)$ is implied by the theorem, but this would mean that it is forbidden to tend the wounds John suffered in a robbery. Clearly something is wrong.

We should not use Prior's formulation of the paradox to evaluate theories of modal logic, I believe, because the analysis of what it is for something to be forbidden (the "Andersonian simplification") is itself not entirely easy to understand, and therefore not something we should feel compelled to accept. As mentioned above, "β is forbidden" is to be understood, within Prior's discussion, as "necessarily, β is subject to sanction". However, the formula $\Box(\beta \to r)$ does not exactly express this. The formula r is a proposition, and so it cannot be paraphrased with a predicate like "is subject to sanction." It must mean something like "sanction is warranted." In that case, $\Box(p \to r)$ means that, necessarily, if someone tends the wounds John suffered in a robbery, sanction is warranted. There is nothing in this formula to say that it is the person who tends John's wounds who will suffer sanction. Rather, the formula expresses a perfectly acceptable conclusion. If someone tends the wounds John suffered in a robbery, sanction is warranted— specifically, the robber should be sanctioned. For this reason, I am not sure that Prior's version of the paradox is really paradoxical.

Chisholm's paradox provides the most compelling illustration of the problems faced by the standard analysis of deontic modality within modal logic. What follows is a formulation by Åqvist (1967):

(89)   (i) $\Box\neg$Mary robs John. (Assumption)
       (ii) Mary robs John. (Assumption)
     (iii) Mary robs John $\to$ $\Box$Mary is punished. (Assumption)
     (iv) $\Box(\neg$Mary robs John $\to$ $\neg$Mary is punished) (Assumption)
     (v) $\Box$Mary is punished.
     (vi) $\Box\neg$Mary is punished.
    (vii) $\Box$Mary is punished $\wedge$ $\Box\neg$Mary is punished.

With (i)–(iv) as assumptions, (v)–(vii) follow logically. (To be precise: (v) follows from (ii) and (iii) and Modus Ponens; (vi) follows from K, (i), (iv), Uniform Substitution, and Modus Ponens; (vii) is the conjunction of (v) and (vi).) But we already saw that sentences like (vii) of the form $\Box p \wedge \Box\neg p$ are both intuitively implausible and incompatible with the requirement of seriality in the possible worlds semantics for D.

It is clear at this point that something is wrong with our standard semantics for deontic modality, and we can turn to figuring out what fundamental lessons we should draw for our linguistic analysis of modal expressions. One assumption made by several of the above arguments is that formulas involving the →, such as *Mary robs John* → □*Mary is punished*, accurately represent conditional sentences. This assumption can be rejected. We saw from the beginning that → is a poor approximation of the English *if... then...* However, even though it is certainly correct that a better theory of conditionals is needed, it would not be correct to say that the only error underlying the paradoxes has to to with the analysis of *if–then*. It is also connected to a fundamental fact about the possible worlds semantics for modal logic we are working with. It is important to see why this is so.

The use of a set of accessible worlds only makes a two-way distinction among worlds: the accessible ones vs. the inaccessible ones. But intuitively a crucial feature of the examples above is that they involve three-way distinctions. For example, worlds in which John's wounds are tended are better than worlds in which they are not, but are worse than worlds in which he never was robbed at all. Likewise, worlds in which there is no robbery are the best; worlds in which John is robbed and Mary is punished are less good; and worlds in which John is robbed and Mary is not punished are worst of all. In both cases, we see three groups of worlds (at least): the good, the mediocre, and the bad. With a single accessibility relation, we can draw a line between the good and the mediocre, or between the mediocre and the bad, but we can't do both.

Seen in this way, the primary lesson of the paradoxes is not that we need a better analysis of conditionals. (We do need one, but this isn't the primary lesson.) It's that our modal semantics has no way of representing the fact that what one should do is dependent in complex ways on the circumstances we find ourselves in. For example, if we can achieve it, a world without violent robbery is what we should strive for, but if we cannot, we should try to help the injured. Conditionals are connected to the problem because they allow us to consider, within a single context (or logical proof) both the perfect circumstance with no robbery at all and, hypothetically, the less perfect circumstance where a robbery has occurred.

The paradoxes are also closely related to the problem of graded and comparative modality. Graded and comparative modality show the need to make distinctions in "goodness" within the accessible worlds.

For example, in (84b) we have accessible worlds in which Mary is lost and ones in which she doesn't want to come to the party; the sentence says that the latter are more likely. Similarly, the Samaritan paradox shows the need to make distinctions in "goodness" within the inaccessible worlds. For example, only worlds in which there is no robbery are accessible (this is why *There should be no robbery* is true), but the conditional sentences reveal an intuition that inaccessible worlds in which John is robbed and the robber is punished are better than the inaccessible worlds in which he is robbed but there is no punishment. Fundamentally, what we need within the theory of modality is a system for comparing worlds to one another in terms of how well they measure up to an ideal; once we have that, we will find many uses for it.[7]

Ordering and the conversational background

Comparison and ordering are closely related. For example, it is possible to compare the heights of two people because it is possible to order people according to their heights. Suppose we want to compare the heights of students in the class. One way to begin would be to line everyone up from shortest to tallest; then, we'll say that Mary is taller than John if John is ahead of her in the line. More likely, what we'll do is measure everyone, writing down each student's height in inches or centimeters, and then use the ordering of these measurements, which follows the ordering of numbers, as a proxy for the ordering of heights themselves.

A conversational background can be used to define a partial ordering of worlds.[8] For example, suppose that we have a conversational background g which assigns to world w three propositions, p, q, and r. In Figure 3.2, each circle represents the set of worlds in which the specified propositions are true. Arrows indicate how to travel from worse worlds towards better worlds; for example, the p, q, r worlds (the worlds in which p, q, and r are all true) are the "best" worlds according to this

---

[7] Although Kratzer's papers are the essential references within linguistics, similar ideas have been developed within deontic logic. For example, Hansson (1990, 2001) develops a framework of deontic logic based on preference orderings.

[8] A total ordering is like the ordering of the natural numbers, a reflexive, anti-symmetric, transitive, total relation. The fact that the relation is total means that every two things are ordered with respect to one another; for every a and b, either $a \leq b$ or $b \leq a$. A partial order allows pairs of things which are not ordered with respect to one another; for example, we may have $a \nleq b$ and $b \nleq a$. Thus, a partial ordering is a reflexive, antisymmetric, transitive relation. A conversational background in general only provides a partial ordering, and, as we'll see below, this leads to some complexity in the definitions of modal operators.

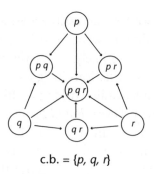

c.b. = {p, q, r}

FIGURE 3.2. A conversational background generates an ordering

conversational background. The p, q worlds, the q, r worlds, and the p, r worlds are less good than the p, q, r worlds, but better than all the others. These sets of worlds in which two of the propositions are true cannot be compared among themselves; that is, a p, q world is neither better nor worse than a q, r world. They are incomparable. The p worlds, q worlds, and r worlds are ordered as indicated in the diagram; for example p worlds are worse than p, q worlds, p, r worlds, and p, q, r worlds, but not comparable with q worlds, r worlds or q, r worlds. Lastly, worlds in which none of the three propositions are true (these are not indicated in the diagram) are worse than all of the others.

We indicate the order generated by a set of propositions X as $\leq_X$. Therefore, if g is a conversational background, $\leq_{g(w)}$ is an ordering generated by the set of propositions $g(w)$. Suppose that g is a deontic conversational background, and $g(w)$ is a set of laws $\{p, q, r\}$ which apply in world w. Then Figure 3.2 indicates that the worlds in which all three laws are satisfied are the ideal ones from the point of view of the laws of w, while worlds in which two of the laws are satisfied are less good than the ideal ones but not the worst; and so forth.

Here's the definition:

(90)   For any set of propositions X and any worlds w, v: $w \leq_X v$ iff for all $p \in S$, if $v \in p$, then $w \in p$.

According to this definition, $w \leq_X v$ means that any proposition in X true in v is also true in w. Thus w is either more highly ranked or ranked the same as v.[9] Because $w \leq_S w$, the order is reflexive. We can

---

[9] Sometimes the positions of w and v on either side of $\leq_S$ are reversed. The idea of having "better" worlds on the left, as we do in (90), is that they line up like the natural numbers, and first is best, like in a race. The idea of having the "better" worlds on the right plays off of the familiar use of $\leq$ to mean "less than or equal to"; that is, if w is ranked higher than v, then more S-propositions are true in w than v.

also make the order irreflexive (written with $<$ instead of $\leq$) by adding to the definition:

(91)   For any set of propositions X and any worlds w, v: $w <_X v$ iff

    (i)   for all $p \in X$, if $v \in p$, then $w \in p$; and
    (ii)   for some $q \in X$, $w \in q$ and $v \notin q$

The ordering in Figure 3.2 is easy to visualize, but depending on the propositions in X, $\leq_X$ can be more complex. It could be that there are propositions in X which are incompatible with each other, such as two laws which can't both be followed. In that case, there can be multiple sets of "best" worlds. For example, suppose that $g(w) = \{p, q, r\}$ and p and q are incompatible. Let's use the notation w_{pr} to indicate any world in which p and r are true but q is false; w_- is any world in which p, q and r are all false. Then we we can describe the order $\leq_{g(w)}$ as follows:

(92)   • $w_{pr} \leq_{g(w)} w_p$
    • $w_{pr} \leq_{g(w)} w_r$
    • $w_{pr} \leq_{g(w)} w_-$
    • $w_{qr} \leq_{g(w)} w_q$
    • $w_{qr} \leq_{g(w)} w_r$
    • $w_{qr} \leq_{g(w)} w_-$
    • $w_p \leq_{g(w)} w_-$
    • $w_q \leq_{g(w)} w_-$
    • $w_r \leq_{g(w)} w_-$

Notice that we not have either $w_{pr} \leq_{g(w)} w_{qr}$ or $w_{qr} \leq_{g(w)} w_{pr}$. There are two incomparable sets of "best" worlds.

Another kind of complexity we can get in the ordering arises when X is infinite in a way which means that there are no "best" worlds; rather, for every world, there is a still better world. For example, Midas always wants to be richer. No matter how rich he imagines himself to be, he prefers to be even richer. His preferences can be represented with a bouletic conversational background which has the following infinite set of propositions in it:

(93) $g(w) =$

    • $p_1 =$ "I have at least one gold coin in the treasure room."
    • $p_2 =$ "I have at least two gold coins in the treasure room."
    • $p_3 =$ "I have at least three gold coins in the treasure room."

- ...
- $p_{1,000,000}$ = "I have at least one million gold coins in the treasure room."
- ...

The ordering has worlds in which he has no gold coins as the worst, those in which he has one are better, those in which he has two still better, and so forth; there are no best worlds because even if he has a million gold coins, a world in which has has a million and one is even better. (Assume that he does not want an infinite number of gold coins, because that would be greedy.) The ordering can be represented as follows, where "better" worlds are towards the left:[10]

$$(94) \quad \cdots \leq_{g(w)} w_{p_{1,000,000}} \leq_{g(w)} \cdots \leq_{g(w)} w_{p_2} \leq_{g(w)} w_{p_1} \leq_{g(w)} w_-$$

In (94), the groups of worlds do not line up like the natural numbers. There's nothing to play the role of 1, and this reflects the fact that there are no "best" worlds.

If we assume that the orderings used by natural language always have a "best" set—that is, no orderings like (94)—we are making what is known as the LIMIT ASSUMPTION. The question of whether we should accept the limit assumption has been discussed by Lewis (1973), Pollock (1976), Herzberger (1979), Warmbrod (1982), and Stalnaker (1987), among others. The dispute usually is framed in terms of the semantics for counterfactual conditionals, and concerns the choice between the closely related theories of counterfactuals put forth by Lewis, who's against the limit assumption, and Stalnaker, who's for it. See Section 5.2 for background and the references cited for details. The argument for the limit assumption is essentially that a situation like that in (94) can never arise in a case of practical interest. To the extent that these arguments are based on special properties of counterfactuals, we should be careful about extending any conclusion we reach about whether to accept the limit assumption in the analysis of counterfactuals to a general conclusion about ordering semantics. That is, it may be that an ordering like (94) can never truly be relevant for the semantic analysis of a counterfactual, but can arise for the semantic analysis of a modal expression interpreted bouletically or in some other way.

---

[10] In this example, p_{n+1} entails p_n. For example, if p_2 is true in a world, so is p_1. Therefore, where I write w_{p_2}, I should really write $w_{p_1 p_2}$, and likewise for all the worlds in which more than one of the propositions are true. But then it would be hard to write $w_{p_{1,000,000}}$ correctly.

Linguists who use a version of Kratzer's ordering semantics for modals often accept the limit assumption (e.g. Portner 1998; von Fintel 1999) for practical reasons. The first such reason is simplicity: with the limit assumption, the meanings of modals are much less complex. The limit assumption is also attractive because it puts those meanings into a format familiar from modal logic and because it renders them parallel to other kinds of quantification, for example nominal quantifiers of the form *every N*. Kratzer (1981) does not make the limit assumption, and in the rest of this section, we will not either. However, whenever possible I will explain the meanings of modal expressions both as they would be expressed following the limit assumption, since this will be easier for first-time students of modality to understand, and as they are actually given in Kratzer's paper.

The semantics of modal expressions in ordering semantics

In the system of Kratzer (1981), a sentence is interpreted with respect to two conversational backgrounds: $[\![\ ]\!]^{c,f,g}$. The first one, f, is called the MODAL BASE, and is used to provide a set of relevant worlds: the worlds in $\bigcap f(w)$. The second one, g, is called the ORDERING SOURCE and it orders the worlds as $\leq_{g(w)}$. If we make the limit assumption, we can say that the set of worlds in $\bigcap f(w)$ which are best-ranked according to $\leq_{g(w)}$ are the accessible worlds for a simple modal sentence. That is, the best-ranked relevant worlds are accessible. The result is (95):

(95) If N is a necessity modal, $[\![\ N\beta\ ]\!]^{c,f,g} = \{w : \{v : v \in \bigcap f(w)$
and there is no $v' \in \bigcap f(w)$ such that $v' \leq_{g(w)} v\} \subseteq [\![\ \beta\ ]\!]^{c,f,g}\}$

The set $\{v : v \in \bigcap f(w)$ and there is no $v' \in \bigcap f(w)$ such that $v' \leq_{g(w)} v\}$ is the set of worlds in $\bigcap f(w)$ which are "best" according to $\leq_{g(w)}$, and so an abbreviated way of writing this is:

(96) $[\![\ N\beta\ ]\!]^{c,f,g} = \{w : \text{BEST}(f(w), g(w)) \subseteq [\![\ \beta\ ]\!]^{c,f,g}\}$

It may be helpful to relate this meaning to the standard treatment in modal logic. BEST plus f and g are in effect combined to create an accessibility relation. In fact, we can define the accessibility relation $R_{f,g}$ corresponding to a modal base f and an ordering source g as follows: $R_{f,g}(w, w')$ iff $w' \in \text{BEST}(f(w), g(w))$. Based on $R_{f,g}$, we could present the meaning of a necessity modal in a format very close to that of modal logic, as follows:

(97) $[\![\ N\beta\ ]\!]^{c,f,g} = \{w : \forall w'[R_{f,g}(w, w') \rightarrow w' \in [\![\ \beta\ ]\!]^{c,f,g}]\}$

Without the limit assumption, things are more complicated. Definition (98) gives the central definition of necessity in Kratzer's system. This rule can be cast as a semantic rule for necessity modals as in (99).

(98) **Necessity:** A proposition p is a necessity in w with respect to a modal base f and an ordering source g iff for all $u \in \bigcap f(w)$, there is a $v \in \bigcap f(w)$ such that:

    (i) $v \leq_{g(w)} u$, and

    (ii) for all $z \in \cap f(w)$: If $z \leq_{g(w)} v$, then $z \in p$.

('Human necessity' in Kratzer (1981))

(99) If N is a necessity modal, $[\![\, N\beta \,]\!]^{c,f,g} = \{w : \text{for all } u \in \bigcap f(w),$ there is a $v \in \bigcap f(w)$ such that:

    (i) $v \leq_{g(w)} u$, and

    (ii) for all $z \in \cap f(w)$: If $z \leq_{g(w)} v$, then $z \in [\![\, \beta \,]\!]^{c,f,g}\}$

Definition (99) essentially says that every sequence of comparable worlds eventually reaches a point, as we move towards "ever-better" worlds, at which we only see worlds in which β is true. Figure 3.3 will be helpful in working through this definition. It shows two infinite series of worlds: the series which proceeds vertically $(w, w_c, w_b, w_a, \ldots)$ has w as the lowest-ranked world, then w_c as more highly ranked, w_b as still more highly ranked, and so forth, with no "best" world. That is, the ordering is $\ldots \leq w_a \leq w_b \leq w_c \leq w$, putting better worlds before worse ones. The series which proceeds horizontally $(w, w_z, w_y, w_x \ldots)$ is similar. The worlds in the two series cannot be compared with each other (except for w, which is in both). The indication "β" shows in which worlds the proposition expressed by some sentence β is true.

Let's go through the definition in some detail. Suppose that Figure 3.3 represents the set of worlds in $\bigcap f(w_0)$ as ordered by $\leq_{g(w_0)}$, for some

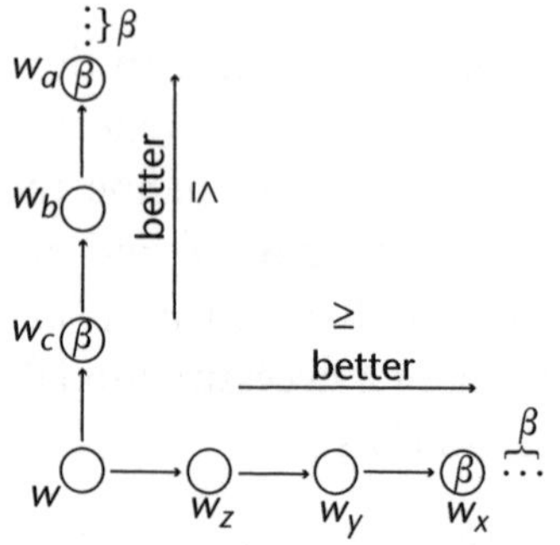

FIGURE 3.3. A complex ordering

particular world w_0. In that case, the proposition defined by (99) is true in w_0. This is so because every world in the diagram meets the description of u in the definition. That is, for each world u in the diagram, there is a v in the diagram such that $v \leq u$ and for all z in the diagram, if $z \leq v$, then β is true in z. For example, consider w_c as on instance of u. We need there to be a v which is (i) as good as or better than u and (ii) for which all worlds which are as good as v or better, β is true. Let's choose w_a as v. It is indeed true that all z such that $z \leq_{g(w_0)} w_a$, β is true in z. Thus w_c meets the characterization of u in the definition. You can check that every other world in the diagram does too. Therefore $[\![\ N\beta\ ]\!]^{c,f,g}$ is true in w_0.

Now we are in a position to examine whether ordering semantics allows an account of graded and comparative modality and whether it provides a solution to the Samaritan paradox. Let's begin with graded and comparative modality. Kratzer (1981, 1991b) defines three notions of possibility:

(100)   **Possibility:** A proposition p is a possibility in w with respect to a modal base f and an ordering source g iff $\neg p$ is not a necessity in w with respect to f and g.

('Human possibility' in Kratzer (1981))

(101)   **Slight possibility:** A proposition p is a slight possibility in w with respect to a modal base f and an ordering source g iff

(i) p is compatible with $f(w)$; and
(ii) $\neg p$ is a necessity in w with respect to f and g.

(102)   **Good possibility:** A proposition p is a good possibility in w with respect to a modal base f and an ordering source g iff there is a world $u \in \bigcap f(w)$ such that for all $v \in f(w)$: If $v \leq_{g(w)} u$, then $v \in p$.

If the ordering source is empty, and so all of the worlds compatible with the modal base are equivalent with respect to the ordering source, then the definitions of human necessity and human possibility become equivalent to necessity and possibility in standard modal logic. Kratzer labels these special cases of human necessity and human possibility "simple necessity" and "simple possibility."

Note that slight possibility amounts to "only a slight possibility" or "no more than a slight possibility." This is so because the definition requires that $\neg p$ be a necessity, in contrast to the definition of (human) possibility in (100), which merely requires that $\neg p$ not be a human necessity. If we make the limit assumption, the distinction can be explained somewhat more concretely: if ϕ is only a slight possibility, it is false is all of the best-ranked relevant worlds. If ϕ is a human possibility, it is true in some best-ranked relevant world. Unlike the other forms of possibility, "slight possibility" is not merely weak in that it only says that ϕ is true in some, rather than all, accessible worlds; it's negative in that it says that ϕ is false in the worlds which conform best to the ideal represented by the ordering source, though true in less-than-best worlds compatible with the modal base.

Good possibility is stronger than (human) possibility because it requires that there is some point in the ordering of worlds where, for all better worlds, p is true. This rules out the situation, for example, where we get an infinitely alternating sequence of p and $\neg p$ worlds. In contrast, this kind of ambivalence between p and $\neg p$ is compatible with human possibility, since human possibility merely requires that $\neg p$ not be a necessity, as it is not in this situation.

Comparative possibility can also be defined as in (103):

(103)   **Better possibility**: A proposition p is a better possibility than a proposition q in a world w, in view of a modal base f and and ordering source g iff

    (i)   For all $u \in \bigcap f(w)$, if $u \in q$, then there is a world $v \in \bigcap f(w)$ such that $v \leq_{g(w)} u$ and $v \in p$, and

    (ii)   There is a world $u \in \bigcap f(w)$ such that $u \in p$ and there is no $v \in \bigcap f(w)$ such that $v \in q$ and $v \leq_{g(w)} u$.

Condition (i) says that, for *every* q-world, there is a p-world which is at least as good. Condition (ii) says that there is a p-world better than all q-worlds. (Condition (ii) is needed to rule out the possibility that the ordering alternates infinitely between p- and q-worlds as in $\ldots \leq w_p \leq w_q \leq w_p \leq w_q \leq w_p \leq w_q$.)

In terms of the definition of being a better possibility we can define weak necessity (Kratzer 1991a):

(104)   **Weak necessity**: A proposition p is a weak necessity in w with respect to a modal base f and an ordering source g iff p is a better possibility than $\neg p$ in w with respect to f and g.

Having developed a range of modal concepts beyond simple necessity and possibility, Kratzer is in a position to give an analysis of graded and comparative modality in natural language. Table 3.1 summarizes her analyses of some expressions of German in Kratzer (1981), along

TABLE 3.1. Force of German modals in Kratzer (1981)

Simple necessity	*Muß*	'must'
Necessity	*Muß*	'must'
	Es ist wahrscheinlich daß	'it is probable that'
Possibility	*Darf*	'may'
	Es kann gut sein daß	'there is a good possibility that'
Slight possibility	*Es besteht eine geringe Möglichkeit daß*	'there is a slight possibility that'
Simple possibility	*Darf*	'may'
Better possibility	*Es kann eher sein daß ... als daß ...*	'it is more likely that ... than that ...'

with translations in English. Table 3.2 gives her analysis of English modals from Kratzer (1991b).[11]

Here is an example showing how the modal force of necessity works for a particular English sentence containing the word *must*:

(105)   (a)   You must have the flu.
     (b) $\{w : \text{for all } u \in \bigcap f(w), \text{there is a } v \in \bigcap f(w) \text{ such that:}$

     (i) $v \leq_{g(w)} u$, and

     (ii)   for all $z \in \cap f(w)$: If $z \leq_{g(w)} v$,

        then $z \in [\![\text{You have the flu}]\!]^{c,f,g}\}$

Imagine a doctor trying to diagnose your illness. The modal base is epistemic, certain facts which are beyond doubt. Among these facts are prominently your symptoms: you have a fever and you have a cough. Other facts could be relevant too: you did not get a flu shot. The ordering source represents expectations which are not beyond doubt, but which the doctor takes into consideration, including his or her training, secondhand reports, experience, and rules of thumb: flu leads to a fever, many people in town are suffering from flu right now, people suffering from the same symptoms in the same town all have the same disease. With this modal base and ordering source, (105b) says that among the worlds in which all of the facts hold, the

---

[11] It is possible that Kratzer would have reconsidered her definitions of the German modals between Kratzer (1981) and Kratzer (1991b) so that they would be more in line with their English translations; however, here I present her analyses of the expressions in German and English as they are found in the articles.

TABLE 3.2. Force of English modals in Kratzer (1991b)

Necessity	*Must*
Weak necessity	*It is probable that*
Possibility	*Might*
Slight possibility	*There is a slight possibility that*
Good possibility	*There is a good possibility that*
Better possibility	*It is more likely that . . . than that . . .*

doctor's expectations rank most highly those worlds in which you have the flu. (The statement is therefore weaker than the unmodalized *You have the flu*, since the propositions in the ordering source are subject to doubt; cf. Karttunen 1972.[12]) It seems to me that the truth of (105a) in this situation is accurately represented by Kratzer's theory.

In the above example, *must* has modal force of necessity, an epistemic modal base and a doxastic (or perhaps stereotypical) ordering source. While it is rigidly associated with the modal force of necessity, other modal bases and ordering sources are possible. When its modal base its epistemic, its ordering source can be doxastic, stereotypical, or empty. Its modal base can also be circumstantial, and if so, its ordering source can be deontic, bouletic, or teleological, or empty. Here are some examples:

(106) (a) The book must have been checked out. (epistemic, doxastic)

     (b) You must turn at the next light. (circumstantial, teleological)

     (c) I must have that painting. (circumstantial, bouletic)

     (d) We all must die. (circumstantial, empty)

Here we see how Kratzer's theory is able to capture subtle differences in meaning and to categorize modals at different levels. For example, (106b–d) are alike in that all have a circumstantial modal base, and in

---

[12] von Fintel and Gillies (2007b) dispute this analysis of the fact that sentences with epistemic *must* seem weaker than the corresponding statements without, saying "We have to dispute the claim that must-claims are weaker than unmodalized claims" (p. 7). They give examples in which a sentence with *must* carries no implication of weakness. However, Kratzer does not claim that sentences with *must* signal weakness or are invariably weaker than ones without. The only fact to be explained is that sentences with *must* are sometimes weaker, and Kratzer's theory makes the correct predictions.

this way they differ as a group from the epistemic (106a).[13] But these three examples differ in the type of conversational background used as the ordering source. Kratzer's work contains many examples exploring the precise meanings which can be expressed by different modals in German and, to a lesser extent, English.

Complex expressions of probability and possibility

One possible objection to Kratzer's treatment of graded modality is that, while it provides a wider range of forces than modal logic, it cannot provide enough. While she provides an analysis of *it is probable that*, natural language has many related expressions including *there is a 60 percent probability that* and *there is a roughly two in three chance that*. It seems that we need a theory which can make very fine distinctions in probability, and it is impossible for Kratzer's theory to make such distinctions. Perhaps this shows that her theory is wrong. As an alternative, we might opt for an analysis of epistemic modality based on probability theory (e.g., Frank 1996; Swanson 2006a, 2006b; Yalcin 2007a). According to such an approach, we measure propositions in terms of their probability, so that $prob(p)$ is the probability of p, and (107a) is analyzed as (b):

(107) (a) There is a 60% probability that it's raining.
       (b) $prob(\{w : \text{it is raining in } w\}) \geq .6$

In considering the importance of examples like (107a), one question is whether complex expressions of probability should in fact be analyzed in terms of the theory of sentential modality. That is, should *there is a 60 percent probability that S* be analyzed in terms of the same theory as accounts for sentences involving *must* and *may*? One could certainly doubt that complex expressions of probability are modal in the relevant, narrow sense. *Must* and *may* are widely attested in human language, and obviously existed before the development of a mathematical understanding of probability; in contrast, *there is a 60 percent probability that* expresses a meaning that had to be invented (or discovered) through the advancement of mathematical knowledge. While this does not absolve the semanticist of the need to provide an adequate analysis of *there is a 60 percent probability that*, perhaps it indicates that the meanings of such expressions are different in kind from those of

---

[13] In fact the circumstantial modals represent the category of "root modality", see Section 4.1.

must and *may*. In other words, it could be that *must* and *may* should
be analyzed in terms of a non-mathematical theory, while *there is a
60 percent probability that* is to be understood in terms of a separate
theory presupposing an additional modern mathematical apparatus.
This way of thinking would receive support to the extent that Kratzer's
analysis of graded modality can represent all of the "grades" of modality
which are present in non-mathematical language and no more.

Let us suppose for a moment that numerical expressions of probability should not be covered by our core theory of modality. Even if
this is the case, there seem to be more grades of modality than have
been analyzed in terms of Kratzer's theory. For example, her theory
analyzes *it is probable that* as weak necessity. But then what about *it
is extremely probable that*? This seems to be stronger than *it is probable
that* but weaker than *must*. Since Kratzer's theory does not provide a
level of modal force in between these two, her theory seems to be in
jeopardy. Once again, an account based on probability theory seems to
be an attractive alternative.

With regard to all of these examples, an important point to keep in
mind is that their analysis must be compositional. In particular, we
need a compositional analysis of *extremely probable* based on appropriate meanings for the lexical items and the way in which they are
composed. Since *probable* is an adjective and other adjectives can be
modified in a similar way (e.g., *extremely tall*), we should apply a general theory of adjective semantics to this case. The same point goes for
similar modifications involving other modal words (e.g., *there is a good
possibility that*), as well as for the comparative forms of modal adjectives
(e.g., *more probable that... than that...*).

Currently, the most well-supported theory of adjective semantics
builds on the notion of a scale. For example, we can think of the
meaning of *tall* in terms of a scale of height. Formally, a scale is a triple
$\langle S, \prec, D \rangle$ consisting of a set of degrees S, an order $\prec$, and a dimension
D which forms the basis for comparing or taking measurements in
terms of the degrees (see, e.g., Kennedy 2007). An adjective can then be
seen as expressing a measure function, that is a function from objects
to degrees on the appropriate scale. For example, *tall* has a meaning like
the following:

(108)   Define *height* as the function which maps any individual x onto
        the $d \in S_{height}$ such that d is the measurement of x according to
        D_{height}.

(109) $[[\text{ tall }]]^c = [\lambda x . x\text{'s height}]$

The meaning of *tall* in (109) maps each individual to its height, a degree in the scale of height.

Degree terms like *extremely* take the adjective as argument; in the absence of an explicit degree term, Kennedy proposes a null morpheme *pos* which supplies a contextual standard degree $\mathbf{d}_c$.

(110) $[[\text{ pos }]]^c = [\lambda A\lambda x . \mathbf{d}_c \prec A(x)]$

(111)   (a) $[[\text{ pos tall }]]^c = [\lambda x . \mathbf{d}_c \prec height(x)]$
          (b) $[[\text{ Mary is pos tall }]]^c = 1 \text{ iff } \mathbf{d}_c \prec height(\text{Mary}).$

Extremely would be of the same type as *pos*, but would say that $\mathbf{d}_c$ is much smaller than $A(x)$, i.e. that Mary's height is much larger than the standard. (See Klein 1980 for relevant discussion in terms of a different theory of adjectives.)

If we wish to apply this adjectival semantics to *probable* and *possible*, we need an appropriate scale for each adjective. For example, there should be a scale $\langle S_{prob}, \prec_{prob}, D_{prob}\rangle$, where S_{prob} is a set of degrees of probability, $\prec_{prob}$ is an ordering of these degrees, and D_{prob} is the dimension. I think that general theoretical considerations, namely, the need for a compositional analysis and the goal of having a uniform semantics for adjective phrases, forces us to the conclusion that such a scale is necessary to the analysis of the adjective *probable*.

Now we can return to the question of whether complex expressions of probability and possibility allow us to choose between Kratzer's semantics and an analysis based on numerical probabilities. The relevance of a scale seems at first to favor the latter: if we adopt the analysis in terms of numerical probabilities, perhaps we can identify S_{prob} with the numeral probabilities (the real interval $[0, 1]$) and $\prec_{prob}$ with $<$. However, if we examine the facts a little bit more closely, they do not actually favor this kind of analysis. I would like to make two comments in relation to this point:

First, expressions of probability employ a different scale from expressions of possibility. The following data makes this point:

(112)   Probability

          (i)   It is extremely/more/*entirely/*completely probable that . . .
          (ii)   There is a high/higher/large/good/*better/*real probability that . . .

(113)   Possibility

    (i)   It is *extremely/?more/entirely/completely possible that ...

    (ii)   There is a *high/*higher/*large/good/better/real possibility that ...

The general pattern seems to be that we describe the scale of probability in terms of quantifiable concepts like height and size, whereas we describe the scale of possibilities in qualitative terms, for example, as real or complete.[14] This difference is unexpected if the basis for the semantics of both kinds of expressions is numerical probabilities. (We find yet different patterns with other adjectives and nouns, such as *necessary* and *chance*.)

The second comment is that neither the scale of probability nor the scale of possibility seems to be identical to $\langle [0, 1], <, D \rangle$. The interval $[0, 1]$ is closed (it has a maximal member), and so we expect that a degree specification like *completely* to be acceptable. (Compare *completely closed* vs. *completely tall*; the scale for *closed* is closed, since there is a point at which a door can be closed no further, while the scale for *tall* is open.) However, as we see above *probable* cannot be modified by *completely*,[15] and while *possible* can, it doesn't have the right meaning. With the scale based on the probability interval $[0, 1]$, *completely possible* should mean that the probability is 1; however, (114) does not mean that it's certain that it's raining. Rather, it means something like that it's a significant enough possibility that we must take proper account of it.

(114)   It is completely possible that it's raining.

In addition to these considerations pertaining to the analysis of *probable, possibilty*, and the like, there is a further argument that epistemic modal verbs should not be analyzed in terms of numerical probabilities. This argument arises from their relationships with other types of modals. While it makes sense to understand epistemic modals in

---

[14] Kratzer (1991b) assigns meanings to *there is a good possibility that* and *there is a slight possibility that* in terms of her definitions of good and slight possibility. However, these analyses are not compositionally derived from meanings for *possibility* and the modifiers.

[15] One could try to explain the fact that *completely probable* is ungrammatical by proposing that *probable* uses a subpart of the full probability scale. For example, it might use the open interval $(0, 1)$, of which 1 is not a member. However, in that case (i) would presuppose that it's not certain that it's raining, which is incorrect given (ii):

    (i)   It's probable that it's raining.

    (ii)   A: It's certain that it's raining.

        B: No, it's probable, but not certain.

terms of probability, deontic, dynamic, and other modals cannot be analyzed in this way. Because the same words are often used to express epistemic and non-epistemic modality, many theories—including all of those we have considered so far—aim to give a unified analysis of modals. For example, in Kratzer's theory *must* always has the same core meaning, with the difference between epistemic and deontic interpretations reduced to the choice of conversational backgrounds. An analysis of epistemic modality in terms of probabilities precludes such a unified analysis.

Of course, it may be that epistemic and non-epistemic modals should not be given a unified analysis. But even if epistemic and non-epistemic modals are not as close to identical as implied by modal logic and Kratzer's system, it is important to keep in mind that there are links among modal meanings that must be explained one way or another. Not only do the same expressions frequently work in different subtypes of modality, but there are regular patterns of historical change which link the various subtypes of modality (see Section 3.3 below; some recent references are: Traugott and Dasher 2002; Traugott 2006; and de Haan 2006). If all types of modality are analyzed in the same general system, for example, as quantifiers over possible worlds, then historical change can be modeled in a simple way:[16] within such a theory, when a deontic modal acquires an epistemic meaning (for example), it simply acquires compatibility with a new kind of accessibility relation/conversational background. In contrast, if epistemic modals involve probabilities, the change is much more radical. Such an approach will require a more powerful theory of semantic change, and will probably have to explain the fact that the same expressions are often used for epistemic and non-epistemic modality as only the residue of historical change, with no status in synchronic grammar.

It seems to me that complex expressions of probability and possibility do not favor an analysis of epistemic modality based on probability theory. (Mathematically sophisticated language which explicitly builds on our understanding of probability theory would not then be classified as involving epistemic modality, strictly speaking.) But this still leaves us with the question of whether Kratzer's theory is able to account for these expressions. In the way it has been developed up until now, it is

---

[16] This is not to say that the historical explanation is simple; just that it's effect in the semantic system is simple; see the discussion below in Section 3.3.

not. There is no simple way to combine her theory with an analysis of adjective semantics based on scales. But despite the fact that Kratzer's theory as such cannot account for these complex expressions of probability, there may be some not-so-simple way to combine the essential idea of her ordering semantics, namely, the idea that variation in modal force is largely based on the ordering generated by the ordering source, with a scale-based semantics. In order to achieve this, we would need a method of deriving a scale of probability from the ordering source.

Let me conclude this section by sketching a method by which we might use the ordering source, in particular Kratzer's analysis of "better possibility," to generate a scale which can be used in an adjectival semantics for *probable*. Take D_{prob} to be a set of equivalence classes of propositions under the "better possibility" relation and $\prec_{prob}$ to be "worse possibility than" as defined by Kratzer. The adjective *probable* would then map any proposition to the equivalence class of which it is a member. Given such a semantics, (115) would mean that the proposition that it's snowing is a better possibility than the contextual standard:

(115)   It is probable that it's snowing.

This interpretation is not identical to the one which Kratzer's theory assigns it, so further work is needed to decide which is preferable. In addition, we'd need to make progress on several other fronts, including: (i) answering the question of what types of contextual standards are compatible with the use of *probable*; (ii) making explicit the meanings of words that can combine with *probable*, for example, *extremely*; (iii) explaining the relationship between the adjective *probable* and the noun *probability*; and (iv) developing parallel analyses of other relevant lexical items, including *possible* and *necessary*, which explain the differences among these words.

In sum, from the perspective of Kratzer's or a similar approach, the best outcome would be a theory of modality which does the following two things:

(i) It treats modal auxiliaries and other functional modal elements as in the original theory.

(ii) It analyzes lexical, sub-sentential expressions of modality and probability (adjectives, nouns, and the like) with compositional mechanisms which are based on ordering semantics, which are appropriate to these categories, and which achieve meanings equivalent to or better than the original analysis.

Of course, there is no guarantee that the approach sketched here will work, but it should at least give an idea of how one might think about the meanings of complex expressions of probability and possibility.

Weak necessity modals

In addition to the compositionally complex expressions of probability that we have been examining, there is another way in which Kratzer's theory may fail to allow for a great enough variety of modal forces. A paper by von Fintel and Iatridou (2006) discusses the semantics of weak necessity modals like *should* and *ought*. (Note that the term "weak necessity modals" is their way of referring to a particular class of modals, and does not necessarily have anything to do with Kratzer's concept of weak necessity.) Such modals have both non-epistemic and epistemic uses, as in the following (their (3), (8)):

(116)   (a)   You ought to do the dishes but you don't have to. (deontic)
          (b)   Morris ought to be in his office. (epistemic or deontic)

Ought to is clearly weaker than *have to* (as shown by (116a)) and *must*. Within Kratzer's system, there are two ways we could explain this. First, as discussed in Section 2.3.3, all of these modals could be associated with the modal force of necessity, but the accessible worlds for *ought* could be a subset of the accessible worlds for *have to/must*. We could enforce such a relation by the appropriate selection of modal base and ordering source; within Kratzer's framework, it makes more sense to focus on the ordering source, since the modal base is presumably circumstantial in both cases. Or second, *ought* might have a different modal force than *have to/must*, in particular weak necessity. Note that this would lead to the epistemic version of *ought* being equivalent to *probable* (assuming we follow Kratzer's analysis of *probable*), apart from any differences due the choice of ordering source.

Instead of choosing either of these alternatives, von Fintel and Iatridou propose to modify Kratzer's system by adding an additional ordering source.[17] They have two reasons for pursuing this option. On the one hand, they have doubts about Kratzer's definition of weak necessity (and more generally on Kratzer's failure to adopt the limit assumption; see von Fintel and Iatridou 2006: footnote 7), and this rules out the second option. On the other, they identify a cross-linguistic pattern in the expression of meanings like that of *ought*—a

---

[17] Note that Kratzer herself suggests the possibility of multiple ordering sources (Kratzer 1981: 72).

cross-linguistic pattern which they feel can be better explained with the assumption that modals can have two ordering sources. Across a wide range of languages, *ought* is expressed with a strong necessity modal like *must* in combination with both future and past tense morphosyntax. For example (their (9)):

(117) Tha eprepe   na   plinis ta piata ala dhen ise
       FUT must+past PRT wash the dishes but NEG are
       ipexreomenos na   to kanis. (Greek)
       obliged   NA it do
       'You ought to do the dishes but you are not obliged to.'

Because the combination of future and past morphosyntax is often a marker of counterfactuality, von Fintel and Iatridou propose that the meaning of *ought* results from the combination of counterfactuality and (strong) necessity.

The core intuition behind von Fintel and Iatridou's use of two order-ing sources is straightforward: when Kratzer added the ordering source to modal semantics, she was able to define levels of modal force in addi-tion to the standard ones of necessity and possibility; it stands to reason that by adding another ordering source, we could make additional distinctions of force. We can see how the extra ordering source works by thinking about (117). This deontic modal has a circumstantial modal base (the relevant facts include that the dishes are dirty, you are not sick, and so forth) and a deontic modal base (the rules of our home). The rules of our home are not very strict, and so the (strong) necessity statement *You are obliged to do the dishes* is false. However, there are other factors in the situation. I did the dishes yesterday and the day before. In all worlds where fairness and family harmony prevail, today you do the dishes. The idea, then, is that the interests of fairness and family harmony contribute a secondary ordering source. *Ought* uses this ordering source, in addition to the deontic one, to rank the worlds, and makes the necessity statement that in all of the best ranked worlds (where the rules are followed and fairness and harmony prevail), you do the dishes. This analysis seems to capture in an intuitive way the way in which *ought* is weaker than *have to*, *must* or *be obliged to*. Note how similar it is to the more conservative alternative suggested in Section 2.3.3: it restricts the set of worlds accessible by *ought* to a subset of those accessible by *have to*, but does so through an extra ordering source, rather than through the modal base or primary ordering source.

In light of this theory, von Fintel and Iatridou explain the presence of counterfactual morphology in terms of the "promotion" of the secondary ordering source. With *have to*, the secondary ordering source is not relevant. But with *ought*, it has been promoted to relevance. Thus, the counterfactual expresses something like "If the secondary ordering source were taken into consideration, it would be a necessity that P" (p. 21). I don't find this explanation for the use of counterfactual morphology very compelling. The secondary ordering source is clearly relevant—that's why *ought* is used—and so saying "if it were relevant" is strange. Nevertheless, it's interesting to see how a natural extension of Kratzer's theory, the addition of more ordering sources, can yield further distinctions in the strengths of modals.

Conditionals in ordering semantics

Kratzer's theory of modality was designed to allow a natural analysis of conditionals. In particular, her solutions to the paradoxes of deontic logic (to be discussed next) rely on this account of conditional modality. Therefore, in order to understand her theory fully, it is necessary to see what she has to say about conditional sentences which contain a modal. Beyond this, she argues that many conditional sentences which seem to lack modals actually include a covert expression of sentential modality, but I set such examples aside for now. We will discuss conditionals further in Section 5.2.

Kratzer proposes that the function of the *if* clause in a modal sentence is to add the proposition it expresses to the modal base of any modal in its scope. Intuitively speaking, we can say that the *if* clause is treated as a fact for purposes of interpreting the modal. There are at least two mechanisms which could be used to carry out this general idea. On the one hand, the *if* clause could be an argument of the modal; this idea is pursued in Kratzer (1977). On the other, it could affect the modal base through independent semantic or pragmatic principles, as in Kratzer (1981). This latter approach is more convenient here, so suppose we have the following structure:

(118)

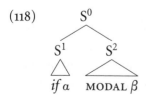

Then the following interpretation rule adds the proposition expressed by a to the modal base for S^2:

(119) $[\![S^0]\!]^{c,f,g} = [\![S^2]\!]^{c,f^+,g}$,
         where for any world w, $f^+(w) = f(w) \cup \{[\![a]\!]^{c,f,g}\}$.

For example, (120a) could be interpreted with respect to an epistemic modal base and a doxastic ordering source. According to rule (119) for conditionals, the meaning of this example is (120b):

(120)   (a)   If it's cold, it is probable that it's snowing.
         (b) $[\![(120a)]\!]^{c,f,g} = [\![$ It is probable that it's snowing $]\!]^{c,f^+,g}$,
               where for any world w, $f^+(w) = f(w) \cup$
               $\{[\![$ It's cold $]\!]^{c,f,g}\}$.

By adding the proposition that it's cold to the epistemic modal base, we in effect treat this proposition as a fact for the purposes for interpreting *it is probable that it's snowing*. Following the definition of weak necessity in (104), the sentence is true iff the proposition that it's snowing is a better possibility than that it's not snowing with respect to f^+ and g. One can cash out the exact truth conditions by applying the definition of better possibility from (103).

Solutions to the paradoxes

Now we return to the paradoxes of deontic logic. They key to solving these problems is the ordering source. Let us begin by looking at the Samaritan paradox as presented by Åqvist, repeated here:

(121)   (a)   It ought to be that John is not injured in a robbery.
         (b)   It ought to be that someone helps John, who was injured in a robbery.

These two sentences are consistent, even though their closest representations in modal logic are not. An important observation about (121b) is that the relative clause *who was injured in a robbery* does not simply contribute to the asserted meaning of the sentence. As mentioned above, it probably contributes a presupposition or conventional implicature. In either case, it is presented as a fact whose truth can be taken for granted.

   These sentences are interpreted with respect to a realistic modal base f and a deontic ordering source g. In the real world w, $f(w)$ contains a set of facts relevant to the interpretation of the modal. Among these facts would be everything presupposed true in the conversation (the common grounds is a subset of $f(w)$, see Portner 2007b), and perhaps other facts as well. What's crucial is that, in the case of (121a), the modal base does not contain or entail the proposition that John was injured in a robbery. To use this sentence with a modal base which entailed that John was injured in a robbery would turn the sentence into a contradiction (it would express the proposition $\emptyset$, true in no world at all), and so language users know to avoid employing such a modal base.

The deontic ordering source assigns to w a set of moral principles. Among these principles will be ones stating that violent robbery does not occur and another stating that we help those who are injured. Because there are worlds in $\bigcap f(w)$ in which John was not injured in a robbery, $\leq_{g(w)}$ will rate any world in which he is not injured in a robbery as better than otherwise similar worlds in which he is. Speaking in terms of the limit assumption, all of the "best" worlds will be ones in which he is not injured in a robbery. Therefore, the sentence (121a) is true in w.

The only difference between (121a) and (121b) is that the latter involves the relative clause which indicates that the proposition that John was injured in a robbery can be taken for granted. Moreover, because the person who said (121b) put this relative clause together in the same sentence with the modal, he or she presumably intends to indicate that it is relevant to the interpretation of the modal statement. Because of this, we can infer that (121b) is interpreted not with respect to f, but with respect to a different modal base f'. The proposition that John was injured in a robbery will be in $f'(w)$, for every w in the domain of f', and therefore every world in $\bigcap f'(w)$ has John being injured in a robbery. Among the worlds in $\bigcap f'(w)$, the deontic ordering $\leq_{g(w)}$ will rate worlds in which someone helps John as better than otherwise similar worlds in which nobody helps him. Thus, (121b) is true. In sum, the examples (121a) and (121b) are interpreted with respect to slightly different modal bases, and they are both true in w.

Now we turn to Chisholm's paradox as given by (89). Here are the assumptions which lead to Chisholm's paradox translated into English:

(122)   (a)   Mary must not rob John.
       (b)   Mary robs John.
       (c)   If Mary robs John, Mary must be punished.
       (d)   If Mary does not rob John, Mary must not be punished.

Again we work with a realistic modal base f and a deontic ordering source g. As we interpret (122a), $f(w)$ cannot entail the proposition that Mary robs John, since if it did, the sentence would be a contradiction. The sentence is true because the ordering source ranks worlds in which Mary does not rob John as better than otherwise similar worlds in which she does. Turning to (122b), here we have a simple statement of fact. We now know that Mary robs John in w. Therefore, $f(w)$ used to interpret (122a) does not contain every piece of relevant factual information; this makes sense, since a person who's assuming that Mary robbed John would never say (122a).[18]

Sentence (122c) is a conditional; it says that *Mary must be punished* is true with respect to the modal base f^+ and the ordering source g. In this case $f^+(w) = f(w) \cup \{[[\textit{Mary robs John}]]^{c,f,g}\}$. All worlds in $\bigcap f^+(w)$ are ones in which Mary robs John. The moral principles in $g(w)$ will rank most highly those worlds in which Mary is punished. Sentence (122d) is also conditional. In this case $f^+(w) = f(w) \cup \{[[\textit{Mary does not rob John}]]^{c,f,g}\}$. All worlds in $\bigcap f^+(w)$ are ones in which Mary does not rob John. The moral principles in $g(w)$ will rank most highly those worlds in which Mary is not punished, since it is better not to punish the innocent.

Chisholm's paradox is that these four sentences, when translated into standard deontic logic, entail that Mary must be punished and that Mary must not be punished. As pointed out above, the status of the English sentences is complex because one would never say (122a) if (122b) is an accepted fact. Therefore, we can describe the situation as follows:

1. If we take (b) as factual, then the proposition that Mary has robbed John is in $f(w)$, for every relevant world w. In that case (a) and (d) are infelicitous, and we are only considering (b,c). These two sentences entail that Mary must be punished. This is so because (c) says that in all of the best world, once we add the supposition that Mary robs John to the modal base, Mary is punished; (b) ensures that the proposition that Mary robs John is taken as true in the conversation. Therefore it is in the modal base. Hence, in all of the best worlds compatible with the modal base, Mary is punished.

---

[18] They might say *Mary should not have robbed John*. This sentence brings time into the picture in a crucial way: we're looking back to a time when Mary had not robbed John, and making a statement based on the facts as they stood then. This kind of explanation is the basis for solutions to the paradoxes which rely on a logic which combines modal and temporal notions (see, for example, Åqvist and Hoepelman 1981; van Eck 1982; and Thomason 2002; Section 5.1 below has further relevant discussion).

2. If we do not assume that Mary has robbed John, we are considering only (a,c,d). These sentences entail that Mary must not be punished. This is so because (a) says that Mary does not rob John in any of the best worlds, and (d) says that in every best world, once we add the assumption that Mary does not rob John, Mary is not punished. Adding this assumption will have no effect, since (a) implies that Mary does not rob John in any of the best worlds even without the assumption. What this case amounts to is a statement that we should not punish the innocent. (Sentence (c) says that if we add the assumption that Mary robs John, she is punished in all of the most highly ranked worlds. This is true as well, but doesn't undermine the overall inference that Mary must not be punished.)

As we see, it is impossible to infer both that Mary must be punished and that she must not be under a single modal base; therefore the paradox dissolves.

Kratzer's theory is quite successful in analyzing the range of phenomena we've discussed so far, absorbing the ideas of modal logic into a framework able to model the wide range of meanings which can be expressed by modal expressions in natural language. Challenges to this theory mainly come from two directions: On the one hand, some scholars dispute the underlying framework of possible worlds semantics inherited from modal logic and argue that it should be replaced by something a bit different (e.g., dynamic logic, see Section 3.2) or very different (e.g., probability theory, see Section 4.2.1, or ideas drawn from cognitive-functional linguistics, see Section 3.3). On the other, linguists have uncovered additional facts about how modality works in natural language, and some of these seem to go beyond what this theory (or traditional modal logic for that matter) can analyze. These phenomena mostly have to do with the interpretation of modals in discourse; they will come up repeatedly in Chapters 4 and 5.

3.2 Modality in dynamic logic

Example (123a) is acceptable, but (123b) is not (from Groenendijk et al. 1996):[19]

---

[19] This same example is used in a number of papers by these authors. I reference Groenendijk et al. (1996) in particular because it is the first one that students of modality should examine.

(123)   (a)   It might be raining outside. . . . It isn't raining outside.
      (b)   It isn't raining outside. . . . ?It might be raining outside.

To be more precise, there is a difference between (123a) and (123b) if they are thought of as sequences of sentences said by the same person and if we assume that the weather has not changed in the meantime. The '. . .' indicates that some time has passed and the speaker has had a chance to gain new information; perhaps the speaker has opened the curtains and looked outside to check the weather. In the case of (123a), the sequence is completely natural, whereas (123b) indicates that the speaker has changed her mind, and is in this sense less natural. Without this opportunity to check the weather (that is, replacing '. . .' with *and*), both sequences are odd.

Within modal logic or Kratzer's semantics, this contrast is not a semantic one. It involves a combination of both semantic and pragmatic factors, in particular our understanding of what modal base is provided by context at different points. For example, Kratzer's approach would say that *might* here is interpreted with respect to an epistemic modal base and either an empty or doxastic ordering source. Let's say that the ordering source is empty for simplicity. Then the first sentence in (123a) says that in some world compatible with the speaker's knowledge it's raining; let's assume that in some other worlds compatible with the speaker's knowledge, it isn't. Then some time passes, represented by '. . .'. The speaker opens the window and sees that it isn't raining; she can now say *it isn't raining outside*. At this point it is no longer compatible with the speaker's knowledge that it's raining.

With (123b) in contrast, Kratzer's theory would describe the situation as follows: The speaker first says that it isn't raining outside. Henceforth, the proposition that it is not raining will be a member of an epistemic modal base.[20] There are no worlds compatible with such a modal base in which it is raining, and adding additional facts to the modal base during the period '. . .' will not change this. Therefore, the sentence *it might be raining outside*, said by this speaker, has to be false. It can only true if the proposition that it's raining is removed from the modal base, which amounts to a change in opinion or weather.

---

[20] Actually the speaker may be wrong about whether it's raining. Therefore, it is more accurate to say that we can conclude that the speaker thinks she knows it isn't raining, and thus that any statement he makes using an epistemic modal should be understood under the assumption that the proposition that it isn't raining is part of the modal base. This rather complicated way of describing things is much simplified from the perspective of dynamic semantics.

Scholars working in the field of dynamic semantics have sought to provide a purely semantic analysis of the contrast in (123a–b) and a variety of other related facts concerning epistemic modals.[21] More precisely, they have sought to provide a logical analysis of such facts, developing a logical system, dynamic logic, within which some important properties of natural language can be modeled. For example, the contrast in (123a–b) is modeled by dynamic logic because the first formula below is consistent while the second is not:

(124)   (a)   Consistent in dynamic logic: $\Diamond p \wedge \neg p$
         (b)   Inconsistent in dynamic logic: $\neg p \wedge \Diamond p$

In order to give a semantic, logical analysis of this pattern, dynamic logic gives up on the idea that all meaning can be represented in terms of truth conditions. We'll see what dynamic logic uses in the place of truth conditions in the next few sections.

3.2.1 The dynamic view of meaning

Dynamic semantics is one of several theories which share the dynamic view of meaning. Traditional semantic theories think about meaning as a "thing," and abstract object of some sort. Each sentence has its meaning, typically represented in terms of truth conditions. For example, the semantics for modal logic assigns a truth value as the denotation of a sentence; Kratzer's theory determines a proposition, the set of possible worlds at which the sentence is true, for each sentence. The dynamic view, in contrast, thinks of meaning as a kind of action. The meaning of a sentence is its capacity to change the context in which it is used. For example, if I say *It is raining outside,* I may well change your beliefs about what I believe, and may even change your beliefs about the weather. According to the dynamic view, the meaning of the sentence is its potential to produce such changes. An apt metaphor (which was actually inspirational in the development of the theory) is that of a computer program. Running a program will produce some changes in the state of the computer each time it is run; these changes are known as its "output." Which changes it produces in a particular case depend on the initial state of the computer, the "input." The program itself has a general capacity to produce certain changes from input to output. In

---

[21] Dynamic semantics was initially developed with the goal of analyzing indefinite noun phrases and anaphora, and has been applied to issues in presupposition more generally. See, for example, Groenendijk and Stokhof (1990); Groenendijk and Stokhof (1991); Dekker (1993); Chierchia (1995a); Beaver (2001).

other words, the program determines a function from inputs to outputs (if the input uniquely determines the output) or a relation between inputs and outputs (if there are several possible outputs for a given input). If a sentence is a kind of program, it can be seen as determining a function[22] from inputs to outputs. This function, known as a CONTEXT CHANGE POTENTIAL or INFORMATION CHANGE POTENTIAL, may be thought of as the meaning of the sentence.

Of course, the idea that the use of a sentence can produce changes in the world is familiar from speech-act theory (e.g. Austin 1962; Searle 1969), but speech-act theory and the dynamic view of meaning focus on quite different kinds of changes. For example, in describing the illocutionary force associated with the sentence *I order you to leave!* within speech-act theory, it is necessary to take into account the social relationship between speaker and addressee (the speaker must be in a position to give an order, or at least they must take him to be in such a position), the capacities of the addressee (addressee must have the ability to carry out the order), and so forth; it turns out that the description of any speech act is a complex thing. In contrast, theories which adopt the dynamic view of meaning are much more narrowly focused.

As a slogan, we say that, in dynamic logic, the meaning of a sentence is a function which modifies the context in which it's used. But what do we mean by "the context"? The notion of context relevant here is neither the very narrow one we employed in analyzing the indexicality of modals (identifying a speaker, addressee, time of utterance, and so forth) nor a broad one involving all of the aspects of the world relevant to speech acts. Rather, within dynamic semantics the meaning of a sentence can be thought of as something which updates one of two things:

(i) The knowledge of some participant in the conversation; or
(ii) The pragmatic presuppositions of the conversation.

The first perspective (i) is easier to talk about and is the one adopted by the initial works on modality in dynamic logic. According to this way of thinking about dynamic semantics, there is an INFORMATION STATE s corresponding to what somebody knows.[23] As each sentence

---

[22] Or perhaps a relation, though it's usually assumed that a function is appropriate.

[23] Footnote 3 from Veltman (1996):

I use the phrases "knowledge" and "knowledge state" where the reader might prefer "beliefs" and "belief state." Actually, I want the information states s to represent something

ϕ is uttered, it updates the hearer's knowledge: the information state s representing the hearer's knowledge is mapped onto a new state s'. The information change potential of ϕ is represented as $[\phi]$ and the UPDATE of s to s' is represented as $s[\phi] = s'$. (Since $[\phi]$ is a function mapping s to s' it would follow the usual convention to write $[\phi]$ before s, as in $[\phi](s) = s'$, but the dynamic semanticists turn things around because it turns out to be typographically convenient.) The new s' represents what the hearer knows after understanding ϕ. Just below, we'll see in detail how this works in dynamic logic.

The second perspective (ii) is represented by a body of important work within the dynamic approach which preceded dynamic logic, in particular Stalnaker (1974, 1978). Instead of focusing on someone's knowledge, Stalnaker is primarily interested in the effect that the utterance of a sentence has on the information which is shared by the participants in a conversation. At any point in a conversation, the participants in the conversation share certain information. For example, we have been talking about the moon. We have mentioned both that it orbits the earth and that it is far away. We have also discussed whether it is inhabited by moon people, and have agreed that it is. I said that it is as big as a bus and, while you do not believe this, you are not disputing it either, to avoid hurting my feelings. So we have four propositions that are shared: that the moon orbits the earth, that it is far away, that it is inhabited by moon people, and that it is as big as a bus. They are shared in the sense that we mutually accept or PRAGMATICALLY PRESUPPOSE them. This means that we each accept them, for purposes of the conversation; we each are aware that the other accepts them; we each are aware that the other is aware that we accept them, and so forth. Notice that our mutually accepting a proposition does not imply that it is true or even that we both really believe it; for example, you do not believe that the moon is as big as a bus, but you pretend that you do, and it is therefore pragmatically presupposed in our conversation.

The set of mutually accepted, or pragmatically presupposed propositions is the COMMON GROUND. If we assume that a proposition is a set of possible worlds, we can intersect the set of the propositions in the common ground to get a proposition: For any common ground cg, $\bigcap cg$ is the set of worlds which are compatible with all of the shared propositions. In terms of our example, any world in $\bigcap cg$ is one in

---

in between: if s is the state of a given agent, it should stand for what the agent regards as his or her knowledge. Things the agent would qualify as mere beliefs do not count. But it might very well be that something the agent takes as known is in fact false.

which the moon orbits the earth, is far away, is inhabited by moon people, and is as big as a bus. The set $\bigcap cg$ is called the CONTEXT SET. The context change potential of a sentence ϕ, $[\phi]$, updates a context set cs to a new context set cs': $cs[\phi] = cs'$.[24]

What is the relation between these two versions of the dynamic view of meaning? One says that a sentence meaning updates an information state s representing someone's knowledge and the other says that it updates a context set cs representing mutually accepted propositions. One possibility is that s and cs are equally important and equally fundamental. Even if this is so, there might be a single notion of "change potential" covering both information change potential and context change potential, so that a uniform semantics is possible, even if a uniform pragmatics is not. Another possibility is that one notion is basic, and the other of lesser importance. For example, it could be that context change potential is the fundamental notion: what a sentence fundamentally does is change the shared information in a conversation; the hearer might or might not believe any of this shared information. Or it could be that information change potential is the fundamental notion: when someone speaks, he or she trying to change update the addressee's knowledge; shared information is understood in terms of what each participant knows about the other. For example, suppose I know it's raining, and you know it's raining, and I know you know it's raining, and you know I know it's raining, and so forth; in that case, the fact that it's raining can be counted as shared knowledge.

Most semanticists who follow the dynamic view of meaning, and many philosophers as well, assume that context change potential is basic. That is, the fundamental function of a sentence is to update the conversational context. However, there is a close connection between the concepts of conversational context and information state. Each is a proposition, i.e. a set of possible worlds. As a result, though we may assume that the meaning of a sentence is fundamentally a context change potential, it can also be used derivatively as an information change potential. That is, since cg and s are the same kind of thing, propositions, if $cg[\phi]$ makes sense, so does $s[\phi]$.

For purposes of understanding the dynamic semantics of modality, it is not too important whether we follow the context change or information change perspective. While there is obviously a real difference

---

[24] We could also think of a "common ground change potential" if we like, $cg[[\phi]] = cg'$. If we do so, we will define $[\phi]$ in terms of $[[\phi]]$. If $cs = \bigcap cg$, then $cs[\phi] = \bigcap(cg[[\phi]])$.

between a theory of mental attitudes like belief and knowledge and a theory of the conversational context, the formal models of the two things are so similar that a specific linguistic proposal laid out in terms of one approach can typically be translated into the other. Thus, when we learn in the next section about the dynamic version of the modal operator $\diamond$, we can describe its effects either in terms of a conversational context or in terms of an individual's knowledge.

3.2.2 Proto-dynamic logic

Research on modality within the tradition of dynamic semantics typically takes the perspective that the meaning of a sentence is an information change potential. Recall that an information state is simply as a set of possible worlds, i.e. a proposition. For example, the set of worlds in which Ben likes penguins represents the information that he likes penguins. If this information state represents the total of all your knowledge, all and only worlds in which Ben likes penguins could be real, as far as your knowledge goes.

The meaning of a sentence is its potential to update an information state. It is a function from propositions to propositions. For the simplest sentences, those of propositional logic (we can think of these as the sentences of **MLL** lacking modal operators), it is easy to define the function. In order to do so, we need a model $M = \langle W, V \rangle$ consisting of a set of worlds and a valuation providing meanings for the non-logical vocabulary of the language. Notice that we don't need an accessibility relation, because we are not dealing with modals yet, and in fact, the dynamic logic theory of epistemic modality doesn't use accessibility relations.

(125)   Given a model $M = \langle W, V \rangle$ and an information state $s \subseteq W$, the information change potential $[\alpha]^M$ of α is defined as follows:

  1.   If α is an atomic sentence, for any information state s, $s[\alpha]^M = s \cap V(\alpha)$.
  2.   If α is of the form $(\beta \wedge \gamma)$, $s[\alpha]^M = (s[\beta]^M)[\gamma]^M$.
  3.   If α is of the form $(\neg\beta)$, $s[\alpha]^M = s - (s[\beta]^M)$.[25]

We define $\vee$ and $\rightarrow$ as before. Here's an example:

(126)   (i) $s[p \rightarrow q]^M =$
       (ii) $s[\neg(p \wedge \neg q)]^M =$

***

[25] $s - (s[\beta]^M)$ is everything in s which is not in $s[\beta]^M$, in other words the complement of $s[\beta]^M$ with respect to s.

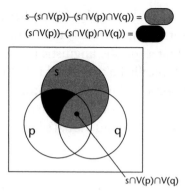

$s-(s\cap V(p))-(s\cap V(p)\cap V(q)) =$

$(s\cap V(p))-(s\cap V(p)\cap V(q)) =$

s

p q

$s\cap V(p)\cap V(q)$

FIGURE 3.4. The result of computing $s[p \to q]^M$

(iii) $s - (s[p \wedge \neg q]^M) =$
(iv) $s - (s[p]^M[\neg q]^M) =$
(v) $s - ((s \cap V(p))[\neg q]^M) =$
(vi) $s - ((s \cap V(p)) - (s \cap V(p))[q]^M) =$
(vii) $s - ((s \cap V(p)) - (s \cap V(p) \cap V(q))$

The gray area in Figure 3.4 shows the information state that results from computing $s[p \to q]^M$. It is s minus the black area. The black area is the portion of s in which p is true and q is false. So $s[p \to q]^M$ is all of s except the portion in which p is true and q is false.

Here's another way to arrive at the same set of worlds. Find the worlds which make $(p \to q)$ true in the non-dynamic logic defined in Section 2.2 (this is everything except the portion of p which is not in q) and then intersect this set with s. The result is again the gray area. In other words, we would get the same result by computing $s \cap \{w : [\![(p \to q)]\!]^{w,M} = 1\}$. In general, the system in (125) has a close relation to non-dynamic logic; this relation allows us to define a dynamic meaning $s[\phi]^N$ in terms of a non-dynamic meaning $[\![\phi]\!]^{w,M}$ as in (127).

(127)   For any sentence ϕ of **MLL** which does not contain a modal operator, $s[\phi]^N = s \cap \{w \in W : [\![\phi]\!]^{w,M} = 1\}$
        (where the models N and M share the same set of worlds W and valuation function V).

A related point is that the set of worlds in which ϕ is true in non-dynamic logic, $\{w \in W : [\![\phi]\!]^{w,M} = 1\}$ can be computed in dynamic logic as $W[\phi]^M$. In other words, $s[\phi]^M = s \cap W[\phi]^M$.

The equivalence (127) makes clear that, in the proto-dynamic system, the contribution of ϕ to the ongoing discourse can be factored into a general principle ("intersect s with...") and something which is not dynamic ("…the proposition expressed by ϕ"). In other words, the dynamic logic we have so far is not essentially dynamic; in can be defined in terms of a non-dynamic logic. I will call it "proto-dynamic." More formally, we can achieve the same results as our proto-dynamic logic by having a traditional static (non-dynamic) semantics, and then employing the following general rule:

(128)   The information change potential of a sentence ϕ is that function $[\phi]$ from information states to information states such that, for any information state s and model M,

$$s[\phi] = s \cap \{w \in W : [\![\phi]\!]^{w,M} = 1\}$$

Some semanticists actually think it is good to be able to have a general rule of this sort. For example, we might believe that sentences contain an force-indicating element (e.g., ASSERT for declaratives) which gives the sentence its ability to update the context (for example, see Krifka 2001; Portner 2004; McClosky 2006, for relevant discussion). The force indicator boosts the non-dynamic meaning into a dynamic one using the relation in (128). However, real dynamic logic is essentially dynamic in that, as other operators are added, they produce a system which cannot be reduced to a static semantics plus a rule of this sort. In particular, the interpretation rules for $\Diamond$ and $\exists$ are essentially dynamic. In the next section, we'll examine the dynamic definition of $\Diamond$.[26]

3.2.3 Dynamic modal operators

It is possible to add a traditional frame semantics for the $\Diamond$ operator to proto-dynamic system. To do this, we would use a model that includes an accessibility relation, and define $\Diamond$ as usual:

(129)   Given a model $M = \langle \langle W, R \rangle, V \rangle$ and an information state $s \subseteq W$, the information change potential $[\alpha]^M$ of α is defined as follows:

1–3.   As in (125).
   4.   If α is of the form $\Diamond\beta$, for any information state s, $s[\alpha]^M = s \cap \{w \in W :$ there exists a w' such that $R(w, w')$ and $w' \in W[\beta]^M\}$.

---

[26] The way to keep $\Diamond$ as essentially dynamic, within a grammar which assigns all dynamic function to force indicators, is to make $\Diamond$ itself a force indicator, or at least one component of a force indicator. Such an approach is certainly possible. See Sections 3.3.3 and 4.2.1 for discussion.

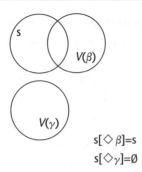

$$s[\lozenge \beta]=s$$
$$s[\lozenge \gamma]=\emptyset$$

FIGURE 3.5. The dynamic modal $\lozenge$

However, this does not produce an essentially dynamic semantics, and it is not the definition of employed within dynamic logic. Instead we have the following rule which does not make use of an accessibility relation:

(130)   Given a model $M = \langle W, V \rangle$ and an information state $s \subseteq W$, the information change potential $[\alpha]^M$ of α is defined as follows:

1–3.   As in (125).
   4.   If α is of the form $\lozenge \beta$, for any information state s, $s[\alpha]^M =$
       (a) s, if $s[\beta]^M \neq \emptyset$;
       (b) $\emptyset$ otherwise.

A sentence $\lozenge \beta$ checks whether β is compatible with s. If it is, that is if $s[\beta]^M$ is not empty, the information state is left unchanged. If it is not, the result is a defective information state, the empty set. Figure 3.5 shows how this works for the simple case where $\lozenge$ combines with an atomic sentence. Suppose that I say to you *It might rain*. If there are worlds in your knowledge state s in which it is raining (*it is raining* is like β in Figure 3.5), s is left unchanged. If there are no worlds in s in which it is raining (*it is raining* is like γ), the logic provides that s is reduced to the defective state $\emptyset$. This would provoke a crisis! So you are not going to let your knowledge state become $\emptyset$. One thing you could do is to reject my statement, refusing to update s with it. Or you might think something is wrong with s, that something you think you know is not really true. As a result, you could revise s—through some mechanism which is not modeled by dynamic logic—to a new state s' which is compatible with the proposition that it is raining.

This information change potential cannot be mimicked using a static semantics plus rule (128). Let β be any sentence which is not necessarily false (that is, assume $\{w : [\![\,\beta\,]\!]^{w,M} = 1\}$ is not empty). We will consider two information states, W (the set of all worlds) and s, where s and β are not compatible (i.e., $\{w : [\![\,\beta\,]\!]^{w,M} = 1\} \cap s = \emptyset$).

1. Under the dynamic interpretation, $W[\Diamond\beta] = W$. To get this result using rule (128), $\{w \in W : [\![\,\Diamond\beta\,]\!]^{w,M} = 1\}$ must be W.
2. Under the dynamic interpretation, $s[\Diamond\beta] = \emptyset$, but if we use rule (128) and if $\{w \in W : [\![\,\Diamond\beta\,]\!]^{w,M} = 1\} = W$, as we just concluded, $s[\Diamond\beta] = s$.

A good way to understanding the dynamic rule for $\Diamond$ is to see how it explains the facts we started this section with, (123), in purely logical terms. Consider the logical analogue of (123a), $\Diamond p \wedge \neg p$. It is possible up begin with a state s, update it with this sentence, and not reach the defective state $\emptyset$. Suppose that both p and $\neg p$ are compatible with s.

(131)   (i) $s[\Diamond p \wedge \neg p]^M =$
         (ii) $s[\Diamond p]^M[\neg p]^M =$
         (iii) $s[\neg p]^M =$
         (iv) $s - s[p]^M$

In contrast, updating s with (123b), $\neg p \wedge \Diamond p$, inevitably results in the empty set.

(132)   (i) $s[\neg p \wedge \Diamond p]^M =$
         (ii) $s[\neg p]^M[\Diamond p]^M =$
         (iii) $(s - s[p]^M)[\Diamond p]^M = \emptyset$

We first compute $s[\neg p]^M$, i.e. $s - s[p]^M$; call this information state s'. Clearly s' contains no worlds in which p is true. Therefore, when compute $s'[\Diamond p]^M$, the result is the empty set. We describe the difference between (123a) and (123b) by saying that the former is CONSISTENT but the latter is not (Groenendijk et al. 1996).

(133)   (a) ϕ is consistent with s iff $s[\phi]^M$ exists and $\neq \emptyset$.
         (b) ϕ is consistent iff there is some state with which it is consistent.

Returning to (123a), we noticed above that while this sentence is consistent, it cannot be uttered by a single speaker "in a single breath."

Groenendijk et al. (1996) explain this in terms of the notions of SUP-PORT and COHERENCE.[27]

(134)   (i) ϕ is supported by s iff $s[\phi]^M$ exists and $= s$.
        (ii) ϕ is coherent iff there is some state which supports it.

A state supports a sentence if the state already contains the information which the sentence provides. According to Groenendijk et al. (1996), we should only say sentences which are supported by our knowledge. If this is right, a sentence can be described as coherent iff it is theoretically possible for someone to say it. The fact that (123a) is not coherent can be seen from the information change potential of its logical representation in (131). Suppose that s is my knowledge state. In order for $s[\Diamond p]^M$ in line (ii) not to be the empty set, s must not support $\neg p$. But since s does not support $\neg p$, it does not support the sentence as a whole (intuitively, I cannot utter the sentence), since, as line (iii) shows, the information change potential of the sentence as a whole is equal to that of $s[\neg p]^M$. This reasoning does not depend on the particular knowl-edge represented by state s, and thus there is no state which supports $\Diamond p \wedge \neg p$; in other words, it is not coherent.

The $\Box$ operator is defined in the usual way as $\neg \Diamond \neg$. Let's run through the calculation, assuming ϕ is an atomic sentence for simplic-ity:

(135)   (i) $s[\Box\phi]^M =$
        (ii) $s[\neg\Diamond\neg\phi]^M =$
        (iii) $s - s[\Diamond\neg\phi]^M =$
        (iv)   (a) $s - s$, if $s[\neg\phi]^M \neq \emptyset$;
               (b) $s - \emptyset$, otherwise
        (v)   (a) $\emptyset$, if $s - s[\phi]^M \neq \emptyset$;
               (b) s, otherwise
        (vi)   (a) $\emptyset$, if $s \not\subseteq (s \cap V(\phi))$;
               (b) s, otherwise
        (vii) (a) $\emptyset$, if $s \not\subseteq V(\phi)$;
               (b) s, otherwise

In other words, $\Box\phi$ performs the test: does the current information s already contain the information in ϕ? If it does (the case in (b)), then it leaves s unchanged; if it does not (the case in (a)), it gives the defective

---

[27] The definition of coherence here is simpler than that of Groenendijk et al. (1996) because we have not introduced the techniques necessary for giving a dynamic semantics for ∃.

state $\emptyset$. The dynamic semanticists do not focus as much attention on the semantics of words which might be thought to correspond to $\square$, such as epistemic *must*; however, as we'll see in Section 4.2, *must* does pose problems for non-dynamic theories and the dynamic $\square$ may provide some inspiration as we look for a solution.

Static dynamic semantics

Yalcin (2007a,b) develops an account of epistemic modality which is in many respects a static version of the dynamic analysis just outlined. In particular, he adopts the idea that $\Diamond\phi$ performs a test on an information state, but he treats this information state as a parameter. As a result, a sentence with an epistemic modal has truth conditions. Here's a version of his definition:

(136) $[\![\, \Diamond\phi \,]\!]^i = \{w : i \cap [\![\, \phi \,]\!]^i \neq \emptyset\}$

In this definition, i plays the role of s in the dynamic analysis: $\Diamond\phi$ performs a test on i. Because w abstracts vacuously in (136) (that is, it does not occur after the colon), $[\![\, \Diamond\phi \,]\!]^i$ is either the set of all worlds or the empty set. As such, if one says $\Diamond\phi$, its function cannot be to change the information state. Rather, the usefulness of $\Diamond\phi$ would come in the information it conveys about i. (Thus our conclusion above holds true: $\Diamond$ cannot contribute to truth conditions in a dynamic semantics which uses a single, general assertion operation.)

Yalcin notes that the crucial patterns which motivated dynamic semantics, for example (123a), are observed in embedded contexts as well.

(137)   (a)   *Suppose that it's raining but it might not be.
          (b)   *If it's raining but might not be, we should stay inside.

These facts fall outside of the theory of dynamic semantics as outlined above because that theory does not provide an analysis of embedded occurrences of epistemic modals. But it is too soon to say that such examples are really a problem for dynamic semantics. Other theories within the dynamic tradition, in particular DRT and File Change Semantics (e.g., Kamp 1981; Kamp and Reyle 1993; Heim 1982, 1982), present dynamic analyses of conditionals, sentence embedding verbs, and other constructions. It is likely that such analyses, combined with the dynamic theory of $\Diamond$, would yield the correct results.

Yalcin also notes that the data in (137) are extremely difficult for more standard analyses such as modal logic or Kratzer's theory to account for, because according to such theories *it's raining but it might not be* has perfectly reasonable truth conditions. It is true iff it is in fact raining but the individual doesn't know for sure that it is. Standard theories were able to explain the anomaly of the unembedded case (123) on pragmatic grounds, but the pragmatic account cannot easily be extended to sentence-internal semantics.

In order to explain the patterns in (137), Yalcin must ensure that the information state parameter i is related in the right way to the worlds in the proposition itself. To see this, imagine that we just happen to set i as W, the set of all worlds. Then the denotation of the second conjunct, *it might not be raining*, is W; it doesn't contribute anything to the meaning, and the analysis fails to capture the contradictory nature of the example. Yalcin counts on the embedding verb *suppose* to make sure that i has the right properties.

(138)   Define $S_{w,x}$ as $\{w' : w'$ is compatible with what x supposes in $w\}$

(139) $[\![\, x \text{ supposes } \phi \,]\!]^i = \{w : S_{w,x} \subseteq [\![\, \phi \,]\!]^{S_{w,x}}\}$

Notice that *suppose* shifts the information state parameter of its complement, so that ϕ is interpreted not with respect to i, but rather with respect to $S_{w,x}$. Given this, (137a) has the following, contradictory meaning:

(140) $[\![\, (137a) \,]\!]^i = \{w : S_{w,x} \subseteq R \text{ and } S_{w,x} \nsubseteq R\} = \emptyset$

Yalcin's theory does not provide such a neat explanation of the original unembedded examples in (123). This is so because, in a root context, there is no operator parallel to *suppose* which can control which information state is used to interpret the sentence with an epistemic modal. To account for this case, a special rule for conjunction is needed. In particular, when we have a conjunction or two independent sentences in discourse, we need to make sure that the proposition expressed by the first clause is incorporated into the information state used to interpret the second. This is precisely what the dynamic conjunction rule in (125) does, and Yalcin needs a "static dynamic" version of it.[28] From

---

[28] Portner (1992) develops a framework which, like Yalcin's, makes use of a context parameter to bring dynamic effects into sentence-internal semantics. Portner provides an analysis of many more compositional structures than Yalcin does (but not epistemic modality), and the rule (i) simplified from Portner (1992: 59) would allow Yalcin to handle (123):

(i) $[\![\, \phi \wedge \psi \,]\!]^i = [\![\, \phi \,]\!]^i \cap [\![\, \psi \,]\!]^{i \cap \phi}$

a broader perspective, what's missing from Yalcin's account is a precise and general explanation of what i is and how it is computed.

Yalcin's theory makes several other contributions passed over here. We already mentioned the probabilistic semantics above, and we will return to some other issues in Sections 4.2 and 5.2.2.

3.2.4 Expectation patterns

Dynamic logic gives an elegant account of the facts concerning (123) and a variety of related facts concerning epistemic modals. It is also useful for understanding another topic closely related to modality, the logic of expectation. In linguistic terms, this yields a semantics for sentences containing the words *normally* and *presumably* (Veltman 1996). The analysis of such expressions makes use of ideas closely connected to the ordering semantics discussed in Section 3.1. It is worthwhile discussing this work briefly, in order to gain a deeper understanding of the ways that ordering can play a role in the analysis of modality.

In order to account for expectations, an information state has to do more than determine a set of worlds. It also has to encode the relative normalcy or expectedness of various worlds. So an information state will now be a pair $\sigma = \langle \epsilon, s \rangle$, where s is a proposition as before and ϵ is an EXPECTATION PATTERN, an ordering of worlds. We read $\langle w, v \rangle \in \epsilon$ as saying that w is at least as expected as v. In other words, every expectation which is met by v is also met by w. This situation is more intuitively written as $w \leq_\epsilon v$.

The absolutely normal worlds are denoted by $\mathbf{n}\epsilon$; the most normal worlds in s are denoted by $\mathbf{m}_{\langle \epsilon, s \rangle}$. It is possible to REFINE an expectation pattern by declaring a proposition p to be a new expectation; this is written $\epsilon \circ p$. Here are the definitions:

(141) (a) **Absolutely normal worlds:**
$\mathbf{n}\epsilon = \{w \in W : \text{for all } v \in W, w \leq_\epsilon v\}$
       (b) **Maximally normal worlds in s:**
$\mathbf{m}_{\langle \epsilon, s \rangle} = \{w \in s : \text{there is no } v \in s \text{ such that } v \leq_\epsilon w\}$
       (c) **Refinement of an expectation ϵ pattern with p:**
$\epsilon \circ p = \{\langle w, v \rangle \in \epsilon : \text{if } v \in p, \text{ then } w \in p\}$

Another way to describe the maximally normal worlds uses the vocabulary of ordering semantics. The worlds in $\mathbf{m}_{\langle \epsilon, s \rangle}$ are the "best" worlds in s from the point of view of the ordering ϵ. Since ϵ represents an ordering of expectedness, these best worlds are the most expected ones, or those which are as normal as possible, given the beliefs we have about how the world really is. We can say our current best guess is that the actual world is in $\mathbf{m}_{\langle \epsilon, s \rangle}$, although our knowledge only guarantees that it is in the wider set s.

Given these definitions, we can now examine Veltman's semantics for *normally* and *presumably*. The job of *normally* ϕ is to refine the expectation pattern by adding a new expectation. The role of *presumably* ϕ is similar to $\Box \phi$. It checks whether ϕ is true throughout the maximally normal (in other words, the most expected) worlds in s; if it is not, we get the defective state $\dagger$.[29] In other words, *presumably* ϕ says that our current best guess is that the actual world is one in which ϕ is true.

(142)   For any information state $\sigma = \langle \epsilon, s \rangle$,

    1. $\sigma[\textit{normally } \phi]^M =$

        (a) $\langle \epsilon \circ \{ w : [\![\phi]\!]^{w,M} = 1 \}, s \rangle$, if $\mathbf{n}\epsilon \cap \{ w : [\![\phi]\!]^{w,M} = 1 \} \neq \emptyset$;

        (b) $\dagger$, otherwise.

    2. $\sigma[\textit{presumably } \phi]^M =$

        (a) σ, if $\mathbf{m}_\sigma \cap \{ w : [\![\phi]\!]^{w,M} = 1 \} = \mathbf{m}_\sigma$;

        (b) $\dagger$, otherwise.

The definition of *normally* ϕ uses the condition that $\mathbf{n}\epsilon \cap \{ w : [\![\phi]\!]^{w,M} = 1 \} \neq \emptyset$. This requires that there be some absolutely normal worlds where ϕ is true, since if the requirement is not met, the update results in the defective state $\dagger$. The intuition here is that when we say *normally* ϕ, we don't just mean that ϕ is an expectation given the facts as we know them in s; rather, it means that ϕ is an absolute expectation. It would be possible to define a different way of refining ϵ which adds ϕ as an expectation only if it's compatible with s; this would give a meaning something like *It is now to be expected that ϕ*.

Notice that the definitions in (142) make use of the static interpretation of ϕ, $\{ w : [\![\phi]\!]^{w,M} = 1 \}$. As with the system of Groenendijk et al. (1996), this can be given in terms of the update function, but the definition is more complex now that we have to deal with the expectation pattern.

(143)   The static interpretation of ϕ = the proposition p such that $\langle E, p \rangle = \langle E, W \rangle [\phi]^M$.

---

[29] $\dagger = \langle W \times W, \emptyset \rangle$. Veltman (1996) uses "**1**" for the defective state, but I fear this would be confusing given the use of 1 to mean "true."

This definition only works for sentences which do not contain $\Diamond$, *normally*, or *presumably*, and so Veltman sets up the logic so that we can only have one of these operators, and only as the outermost operator of the sentence.[30]

Here's an example:

(144)   (a)   Normally, it is raining in Holland. So presumably it's raining in Holland.

    (b)   Take some information state $\sigma = \langle \epsilon, s \rangle$:

        (i) $\sigma[normally(r) \wedge presumably(r)]^M =$

        (ii) $\sigma[normally(r)]^M [presumably(r)]^M =$

        (iii)   (a) $\langle \epsilon \circ r, s \rangle [presumably(r)]^M$,   if $\mathbf{n}_\epsilon \cap \{w :$
$[\![\phi]\!]^{w,M} = 1\} \neq \emptyset$;

            (b)   †, otherwise.

            (The condition $\mathbf{n}_\epsilon \cap \{w : [\![\phi]\!]^{w,M} = 1\} \neq \emptyset$ just says that it is compatible with our other expectations that it's raining in Holland, which it certainly is, so we can ignore that part.)

        (iv)   (a) $\langle \epsilon \circ r, s \rangle$,   if $\mathbf{m}_{\langle \epsilon \circ r, s \rangle} \cap \{w : [\![r]\!]^{w,M} = 1\} =$
$\mathbf{m}_{\langle \epsilon \circ r, s \rangle}$;

            (b)   †, otherwise.

Line (iv) says that the new information state will be $\langle \epsilon \circ r, s \rangle$ if it is raining in Holland in every maximally normal world in $\langle \epsilon \circ r, s \rangle$. It may look like the semantics guarantees us that this condition is met, since $\langle \epsilon \circ r, s \rangle$ contains the expectation (that is, r) that it is raining in Holland. However, it could be that there are countervailing expectations (e.g., *It doesn't rain in the summer*) which, given certain facts (*It is now summer*), imply that not every maximally normal world in s is one in which it's raining in Holland. Veltman's theory addresses such more complex cases as well. What we can say here is that, if there is nothing else in the information state relevant to the question of whether it is raining in Holland, the output of this computation will be $\langle \epsilon \circ r, s \rangle$. In such a case, the whole effect of (144a) is the addition of an expectation that it is raining in Holland to the addressee's knowledge.

---

[30] Actually, Veltman keeps the logical system for *might* (or $\Diamond$) separate from that for *normally* and *presumably*, so technically the issue with *might* does not arise.

3.2.5 Evaluation of dynamic semantics for modality

The initial goal of dynamic logic was to provide an analysis of facts which cannot be explained within classical logic, for example, the pattern in (123). Dynamic logicians have been quite successful at explaining the patterns which they identified as in need of such analysis. We have not gone into many of these here—in particular, I've avoided those concerned with anaphora and its interactions with modality—but they are many, and the explanations for them are elegant. Nevertheless, we can ask whether a logical analysis of these facts is called for in the first place. We saw at the beginning of this section that one can explain (123), for example, in terms of Kratzer's theory. It requires the combination of semantics and pragmatics to do so, of course, but in principle there's nothing worse about a combined semantics-pragmatic account compared to a purely semantic, logical account. For this reason, I feel that the right way to see the contribution of dynamic logic is not in its solutions to the empirical problems which initially motivated it, but rather in the conceptual alternative it offers for how to look at modality.

The key difference between the standard possible worlds semantics for modality and the dynamic approach is that within the former, modals contribute to truth conditions, whereas the within the latter, they do not. In dynamic theories, information change potential (or context change potential) is the fundamental way of modeling meaning, and while simple sentences have information change potentials which can be simplified into static truth conditions, not all sentence do. In particular, sentences with *might, must, normally,* and *presumably* are essentially dynamic. We should ask whether there is further evidence for the idea that epistemic modals and expressions of expectation are essentially dynamic, that is, lacking in truth conditions. As we will see in Sections 4.2, there may well be. At that point, it is likely that the insights of dynamic logic will prove useful in new ways.

Another crucial point for evaluating dynamic semantics is that there is no theory of non-epistemic modality within dynamic semantics.[31] This might be so for either of two reasons. On the one hand, it could

---

[31] There is discussion of non-epistemic modality within the more purely logical end of dynamic logic, e.g. van Bentham and Liu (2004); Yamada (2007a,b). These works develop interesting ideas about obligation, commands, and the like, but are not as closely related to linguistic concerns as those of the dynamic semanticists. Mastop (2005) is an important study of imperatives within dynamic semantics. However, there seems to be no linguistically-oriented discussion of dynamic properties of deontic modals, as opposed to the speech acts associated with imperatives.

be that non-epistemic modals are not essentially dynamic. According to this point of view, they are correctly analyzed within traditional modal logic; their semantics is given by rule 4 in (129), with appropriate choices of accessibility relation for various kinds of modality. On the other hand, it could be that a dynamic theory could and should be given, but has not been simply because scholars working within this community have not been interested in, or have not detected, the dynamic aspects of meaning of deontic and other non-epistemic modals.

I tend to believe that the second possibility is correct. One argument for this position comes from imperatives and deontic modals. Imperatives function to place a new requirement on the addressee and have an intuitive connection to deontic modality (see Portner 2004 and Portner 2007b for discussion). In addition, deontic modals may well be essentially dynamic:

(145)   You must leave now.

As pointed out by many authors (see, for example, Lyons 1977 and the works cited there), this sentence does more than report that the addressee's leaving follows from the relevant rules; in some sense, it creates or imposes the requirement that the addressee leave, in the same way that an imperative would. Lyons, and recently Ninan (2005), make this point clearly by pointing out that (145) cannot be followed by . . . *but you won't*. Ninan gives a formal analysis of this phenomenon, proposing that a root sentences containing deontic *must* makes two contributions to the discourse: (i) it expresses a proposition, typically asserted, that an obligation exists, and (ii) it imposes the obligation in question. We will return to this issue in Section 4.3, but for the time being, let's focus on the question of how we could, in principle, add this type of deontic meaning to dynamic logic.

One point to begin with is that we have already learned, in the discussion of Kratzer's theory, that an ordering semantics for deontic modality seems necessary. Therefore, an analysis of deontic modality is likely to have some similarity to the semantics for *normally* and *presumably* in Veltman (1996). So, if we want to give a dynamic semantics for deontic modality, is it more like *normally* or more like *presumably*? If it is like *normally*, it will refine an ordering of worlds; if it is more like *presumably*, it will test whether a condition holds within the current "best" set of worlds.

Let's work through the two candidates. We begin with an information state $\sigma = \langle \rho, s \rangle$, where ρ is a partial ordering of worlds. We have $w \leq_\rho v$ iff every rule which is complied with in v is also complied with in w. (When we consider sentences involving multiple kinds of non-epistemic modality, we'll need multiple orders, but let's just build a simpler system, in the spirit of dynamic logic, for now.) Given this, here are two candidate meanings for deontic operators:

(146) For any information state $\sigma = \langle \rho, s \rangle$,

1. $\sigma[\Box_1 \phi]^M =$
   - (a) $\langle \rho \circ \{w : [\![\, \phi \,]\!]^{w,M} = 1\}, s \rangle$, if $\mathbf{n}_\rho \cap \{w : [\![\, \phi \,]\!]^{w,M} = 1\} \neq \emptyset$;
   - (b) †, otherwise.
2. $\sigma[\Box_2 \phi]^M =$
   - (a) σ, if $\mathbf{m}_\sigma \cap \{w : [\![\, \phi \,]\!]^{w,M} = 1\} = \mathbf{m}_\sigma$;
   - (b) †, otherwise.

Like *normally ϕ*, $\Box_1 \phi$ uses ϕ to refine the ordering. In the case of a deontic interpretation, this amounts to making ϕ into a new requirement. In contrast and like *presumably ϕ*, $\Box_2 \phi$ checks wither ϕ is true among the "best" worlds in the information state.

If imperatives are to be analyzed as dynamic modal operators, something along the lines of $\Box_1$ is called for.[32] Suppose that ρ is an ordering of words reflecting the addressee's commitments. Then, uttering $\Box_1 \phi$ will make ϕ into a new commitment. It is more difficult to say how deontic modals like *must* would fit into this system. If Ninan (2005) is right and *must* has a semantic function similar to that of an imperative, it would have the meaning of $\Box_1$ as well; however, Ninan thinks that *must* has have a traditional truth-conditional semantics. The truth-conditional semantics is not provided by either $\Box_1$ or $\Box_2$, but would rather have to come from a frame-based analysis as in (129). That is, translating Ninan's ideas into dynamic logic makes *must* into a combination of a dynamic modal operator and a standard modal operator.

If Ninan is wrong and *must* has neither an imperative-like function nor truth conditions, $\Box_2$ provides a plausible dynamic semantics. The idea is that (145) checks whether your leaving is implied by the deontic

---

[32] Compare van Rooij (2000). Mastop's (2005) dynamic analysis of imperatives is similar to this, but much more complex. Han (1998, forthcoming) and Portner (2004, 2007b) provide good background on imperatives semantics. See Portner (forthcoming) for comparison of theories of imperatives.

ordering ρ, given the factual knowledge in s. If it is, all is fine and the information state is left unchanged; if it is not, the semantics provides that we get the defective information state †. But of course, as we saw above, no addressee will tolerate having his or her knowledge state reduced to the defective state, and some discussion must ensue. Either the addressee will reject the utterance entirely or will revise his or her information state (in a way not determined by the logic) to one that allows $\square_2\phi$. This amounts to accommodating ϕ as a requirement (or, at least, accommodating something which amounts to ϕ being a requirement relative to the facts in s).

In sum, dynamic logic is a rich source of ideas and tools for analyzing non-truth-conditional aspects of the the semantics of modal expressions. As we'll see below, there is growing evidence that non-truth-conditional meaning is not a minor issue in the analysis of modality, but rather a central feature. Thus dynamic logic may make important contributions as we build a better semantic theory of modality. However, from the linguist's point of view dynamic logic suffers from its status as a logician's pet. It has only been applied to a few modal expressions and to a narrow range of sentence patterns, and so it is unclear whether the subtle distinctions discussed by Kratzer, and the wide range of facts brought up by other linguists (see Chapters 4–5), can be explained within dynamic logic as it has been formulated up to now.

3.3 Modality in cognitive and functional linguistics

In this section, I will discuss work on modality which has been developed within two traditions which are typically seen as being opposed to, and incompatible with, the theories of modality based on possible worlds. These traditions are COGNITIVE and FUNCTIONAL linguistics. Cognitive linguistics is a theory which aims to explain the nature of language in terms of more general properties and capacities of the mind. Functional linguistics is not a particular theory, but rather a general perspective on how linguistic facts are to be explained. Generally speaking, functional linguistics aims to explain the nature of language in terms of the uses to which language may be put, including both information-based and social uses. Cognitive and functional linguistics do not necessarily go together. For example, a functional linguist might believe that our brains have a separate language faculty (so that the properties of language are not due to general capacities of the mind, but rather language specific ones) while also believing that that this faculty has a functional basis (that it has evolved to have the properties it does because they allow it to serve particular functions). From the other side, a cognitive linguist might believe that general properties of the mind explain the nature of language completely, and that there is no need to refer to the uses of language in giving the explanation. Nevertheless, cognitive and functional linguistics often go together; in particular, cognitive linguistics (the younger of the two traditions) often takes a functional perspective.

Much functional work on modality (as on other topics) is descriptive and does not aim to provide a semantic theory, and for this reason it is a bit difficult to incorporate it into the present book, since this book aims to explain and discuss semantic theories of modality. Saying that much functional work on modality does not provide a semantic theory may be seen as contentious and dismissive of the work in question, but it is not intended to be so, and so let me take a moment to explain what I mean by a "theory," in particular a "semantic theory." Also, note that when I say that someone who studies the meanings of modal expressions does not attempt to develop a semantic theory, this does not mean they do not aim to provide a theory of some other related area, for example, a theory of syntax, language change, or language typology. Much functional research on modality is explicitly developed in terms of such theories.

The concept of a semantic theory only makes sense if there are such things as semantic facts, that is, a set of facts which are true because of the meanings of linguistic expressions. For example, most semanticists include among the important semantic facts the fact that a certain sentence is appropriate and true in given situation; that a relation of entailment, synonymy, or contradictoriness holds between two sentences; and the fact that a certain declarative sentence can be an appropriate answer to a particular question. One might have other conceptions of what the important semantics facts are. For example, if one believes that sentences are only unstable abstractions from a more basic type of thing, utterances-in-context, one might think that it doesn't make sense to say that a sentence entails another. The relevant facts would have to do with utterances-in-context, not sentences. Whatever the basic nature of semantic facts is, the key point is that we need to have some. Suppose we do. Then a semantic theory is a system of ideas which predicts those facts, makes additional predictions about what other facts we may discover, and does all of this in a completely mechanical way, that is, without relying on a prior understanding of the language being analyzed or of the theoretical terms involved. Of course, in order to be a candidate for the right theory, it should not make any predictions which we know to be incorrect.

If we are very tough on ourselves, we should admit that there are, in fact, no theories of semantics. No system of ideas has yet been presented which predicts all of those semantic facts that have been discovered and which does so in a way that relies on no prior understanding of the language being analyzed. We can produce theories which aim to predict some range of the semantic facts we know about, and which are completely mechanical within their limited domain. The fragments of Montague Grammar, such as Montague (1973), are examples. There are also theories which are not completely mechanical, but which could be made sufficiently mechanical with a little bit of work. For example, dynamic logic with modality does not give a mechanical semantic theory because it is nowhere made explicit how to relate actual sentences of a natural language to expressions of the logic; however, the dynamic semanticists assume that experts see how to define a mapping from English sentences to sentences of dynamic logic with modality (perhaps building on Montague Grammar as in Groenendijk and Stokhof 1990, for example). But despite these admirable efforts towards precise, mechanical theories, what you can find in the literature today are virtually always just approximations of theories. They are systems of ideas which have been articulated with the goal of one day developing them into complete and precise theories in the ideal sense. I think it's fair to call such systems "theories" too, in a looser sense.

Thus, when I say that much work on modality in functional linguistics does not aim to provide a semantic theory of modality, what I mean is that it is not part of an effort to create a complete, mechanical model which predicts all of the semantic facts about modals we know about and which makes predictions about additional facts we don't. Functional work on modality aims to explain and define features of modal meaning, but in doing so it relies on a prior understanding of the language being analyzed or of other theoretical concepts. In order to see what I mean, let us look at one interesting paper in the functional tradition, Nuyts (2001b). (I choose this paper because it introduces in a clear way a topic I will discuss further in Sections 3.3.3 and 4.2. Other important works on modality from the functional/descriptive tradition written in English include: Coates 1983; Traugott 1989; Traugott and

Dasher 2002; Bybee et al. 1994; the papers in Bybee and Fleischman 1995a; and Nuyts 2001a.)

Nuyts's paper discusses the concept of SUBJECTIVITY in epistemic modal sentences. The concept of subjectivity was introduced by Benveniste (1971) and applied in detail to English modals by Lyons (1977). According to Lyons, some uses of epistemic modal expressions concern the speaker's subjective evaluation of a proposition, whereas others express the objective probability that it is true. Nuyts suggests two ways of clarifying Lyons's definition:

1. **Subjective vs. objective:** The use of a modal expression is objective if it is based on highly reliable evidence, whereas it is subjective if the evidence is less reliable.
2. **Subjective vs. intersubjective:** The use of a modal expression is intersubjective if it is based on evidence which is shared between the speaker and others, whereas it is subjective if the evidence is held by the speaker alone.

One of Nuyts's main claims is that the second characterization better matches the data he considers, namely, corpus data from Dutch and German. (He further argues on the basis of this that subjectivity and intersubjectivity are more closely related to evidentiality than to modality. We'll return to this idea in Section 4.2.2.)

Nuyts's methodology for trying to decide which concept of subjectivity is correct is to first examine a large number of examples from a corpus, decide to what extent they are interpreted subjectively, and then see which concept correctly describes them. For example, he finds that modal auxiliaries are typically neutral on the dimension of subjectivity, while modal adjectives typically mark a high degree of subjectivity or an extreme lack of subjectivity. A subjective example is the following (Nuyts's (7), 2001b: 390):[33]

[33] One troubling aspect of Nuyts's discussion is that the classification of examples as subjective or not subjective is itself completely subjective; that is, he labels examples as subjective if they seem subjective to him. Therefore, the conclusions he draws from this classification may only reflect the judgments he made, and not the patterns which are truly important in language. However, I don't wish to press this criticism too hard, since I have no reason to doubt that he made the right judgments. (As pointed out by Steve Kuhn, perhaps we could take a judgment of "subjective" as on a par with a judgment of "grammatical." Of course, Nuyts might not accept the validity of grammaticality judgments on methodological grounds, as some linguists do not, and it seems quite unlikely that judgments of grammaticality or ungrammaticality should be seen as more suspect than judgments of subjectivity or objectivity.) In any case, the point here is not to evaluate Nuyts's paper per se, but rather to use it to illustrate the nature of functional work on modality.

(147)   Mijn ontstemming werd nog groter toen mijn kameraden even
        later met een zelfvoldane gelaatsuitdrukking terugkeerden, al
        leek het mij niet *onwaarschijnlijk* dat zij maar branieachtig
        deden alsof zij een opwindend avontuur achter de rug
        hadden.

        'My annoyance increased even more when shortly afterwards
        my friends returned with self-satisfied faces, even though it
        seemed not *unlikely* to me that they were only bragging as if
        [they] had an exciting adventure behind them.'

For native speakers, there is a clear intuition that someone saying this
sentence (or its translation in English) would be expressing a subjective
judgment. In contrast, the following example is not subjective (Nuyts's
(5) 2001b: 389–90):

(148)   In der archaischen Zeit hatte die Mauer bereits densel-
        ben beträchtlichen Umfang wie in späterer Zeit. Es ist sehr
        wahrscheinlich, daß dieser älteste Mauerring, der für die dama-
        ligen Verhältnisse und Vorstellungen ungewöhnlich groß war,
        von dem Tyrannen Polykrates errichtet wurde, [...].

        'In the archaic era the wall already had the same considerable
        circumference as in later times. It is very *probable* that this
        oldest circular wall, which was unusually large for the stan-
        dards and conceptions of those days, was erected by the tyrant
        Polykrates.'

After categorizing examples as subjective or not, Nuyts can then try
to figure out which concept of subjectivity is correct. He says that examples like (148) can be construed both as being based on solid evidence
(they are objective) and as being based on shared evidence (they are
intersubjective). However, in examples like (147), the definition of subjectivity which opposes it with intersubjectivity wins out: "the quality of
the evidence probably matters less than the fact that it is evidence only
available to the speaker, not to the hearer" (p. 394). Overall, judgments
of this kind lead him to conclude that the dimension of subjectivity vs.
intersubjectivity is more important than that of subjectivity vs. objectivity in the meaning of modal expressions. (He draws other important
consequences as well; we'll return to these later.)
    I believe that Nuyts's observations are important and that his con-
clusion about the importance of (inter)subjectivity in modal semantics
is probably correct. But it's also clear that he is not aiming in this

paper to produce a theory of modal semantics. In particular, the only semantic predictions made by Nuyts's ideas concern whether speakers in particular contexts will employ expressions of subjective or intersubjective modality. The key concepts of subjectivity and intersubjectivity are explained in terms such as the following:

[O]ne pole [subjectivity] involves the speaker's indication that (s)he alone knows (or has access to) the evidence and draws conclusions from it; the other pole [intersubjectivity] involves his/her indication that the evidence is known to (or accessible by) a larger group of people who share the same conclusion based on it.   (2001b: 393)

In order to see what predictions are made by Nuyts's ideas, one has to already understand such concepts as "for a speaker to indicate something," "evidence," "for someone to know (or have access to) evidence," "for someone to draw conclusions," and "for a group of people to share a conclusion." To the extent that we already have a good understanding of these concepts, Nuyts's proposal has content and can make a contribution to our understanding of modal meaning. Nevertheless, because it is not part of an effort to create a complete, mechanical model which predicts the relevant facts, his ideas do not constitute a theory in the sense discussed above.

The majority of research on modality within functional linguistics does not present a semantic theory in the same sense that Nuyts's paper does not, but this does not mean that it has nothing to contribute to this book. Rather, as we develop and evaluate semantic theories, we must attend to the findings of the functional literature. Indeed, I believe that the concepts of subjectivity and intersubjectivity are quite important and must be incorporated into our analysis of modal meaning, a point we'll return to in Section 3.3.3.

3.3.1 Modality as representing force dynamics

Most functional research does not present a theory of semantics, but one functional approach does: cognitive semantics. Although the ideas of cognitive semantics are not presented in the same rigorous and precise way as theories based on possible worlds, it is clearly the goal of researchers in this tradition to develop proposals which can, in principle, be incorporated into a complete, precise, and mechanical theory of meaning in natural language. In this section, I will explain the ideas about modality which have been developed within cognitive semantics.

As mentioned above, one of the major emphases of cognitive semantics is the idea that meaning is to be explained in terms of general properties of cognition. Cognitive semanticists have offered a number of theses about how this works, and a common theme is that there is a range of basic concepts which are derived from our experiences as living creatures with bodies moving about in space and time, and then metaphorical extensions of these to other, more abstract domains (see, for example, Johnson 1987; Lakoff 1987; Lakoff and Johnson 1980; Fauconnier and Turner 2002).

One fact about the world that we understand well is that objects may be in motion or at rest, and that changes in their being in motion or at rest result from forces acting upon them. That is, we have concepts of FORCE DYNAMICS which we use as we interact with the world of objects and forces (Talmy 1988). Talmy presents a model of force dynamics involving two theoretical entities, the AGONIST, the thing for which the possibility of movement is relevant, and the ANTAGONIST. In the most basic case, the agonist may be associated with a force keeping it in place or a force tending it towards motion. The antagonist exerts a force contrary to that associated with the agonist. Suppose that the agonist is associated with a force tending it towards motion and that the antagonist is associated with an opposing, but weaker, force. This situation would be described with a sentence like the following, Talmy's (3c):

(149)   The ball kept rolling despite the stiff grass.

Translated into the terminology of lay physics, the ball is the agonist, its force is its inertia or the gravity pulling it down the hill, the grass is the antagonist, and its force is that created by friction and the stiffness of the grass. Talmy adds a number of refinements towards this basic system, including the possibility that the relative strengths of the forces change over time or that the antagonist introduces no force at all on the agonist (it is out of the way).

Talmy argues that deontic modal meanings are metaphorical extensions of force-dynamical concepts. In particular, forces are metaphorically mapped onto psychosocial relations like desire and power. For example, consider (150a). According to Talmy, this sentence means "I exert a force of authority causing you to decide to leave."

(150)   (a)   You must leave (because I say so).
         (b)   You may borrow the book.

More abstractly, (150a) describes situations in which the speaker is the antagonist, the referent of the subject (in this case, the addressee) is the agonist, the metaphorical dimension of motion is a change from a plan not to leave to a plan to leave, the speaker's authority is the antagonist's force, and the addresee's desire not to leave is the agonist's force. The use of *must* in (150a) indicates that the antagonist's force is greater than the agonist's. The use of *may* in (150b) indicates that the antagonist introduces no force opposing the agonist at all (though it could have). See Table 3.3 for a summary.

Sweetser (1990) extends Talmy's analysis into the domain of epistemic modality (see also Bybee and Pagliuca 1985). Taking her cue from the historical evidence that epistemic meanings typically derive from non-epistemic ones, she argues that the physical forces of force dynamics and the psychosocial forces which are relevant to deontic modality are mapped onto inferential forces in the case of epistemic modality. For example, (151a) means something like "the evidence exerts a cognitive force causing me to come to believe that it is raining."

(151)   (a)   It must be raining.
        (b)   It may be raining.

In terms of Talmy's force dynamical modal, in (151a–b) the speaker is the agonist, the metaphorical dimension of motion is a change from a lack of belief that it's raining to a belief that it's raining, a body of evidence of which the speaker is aware is the antagonist, and the support which the evidence offers for the change of belief is the force associated with the antagonist. (It's not clear what the metaphorical force associated with the agonist is in this case. Perhaps there is none. Alternatively, in (151a), perhaps it is the desire not to believe that it's raining, or more fundamentally the desire not to believe falsehoods, but in the case of (151b) a desire to believe that it's raining seems more intuitive.) Again, see Table 3.3 for a summary.

Later work on modality in cognitive linguistics has developed the ideas of Talmy and Sweetser in various ways but has not altered the key ideas of the force-dynamics model. One of the most important concerns has been the nature of subjectivity, as in Langacker (1990, 1999) and the articles in Athanasiadou et al. (2006), for example. We will return to this topic in Section 3.3.3.

Now that I have outlined the cognitive semantics view of modality, we can turn to evaluating it as a semantic theory. It is plausible that there is some kind of cognitively real relation—possibly a metaphorical

TABLE 3.3. Examples of force dynamics for deontic and epistemic modality

	Deontic (150)	Epistemic (151)
Agonist	Subject (addressee)	Speaker
Agonist's force	Subject's desires	None (?)
Antagonist	Speaker	Evidence
Antagonist's force	Social authority	Weight of evidence
Metaphorical motion	Towards decision to leave/borrow book	Towards belief that it's raining
Meaning in (a): *must*	Social authority stronger than desire	Weight of evidence sufficient to compel belief
Meaning in (b): *may*	Social authority does not oppose desire	Weight of evidence does not prevent belief

one—between the basic domain of force dynamics and the concepts which allow our understanding of modality. But as we evaluate the ideas above for their ability to provide a theory of the meanings of modal expressions, we must ask for more than that. The theory must explain what modal meaning *is*, not just what it is *like*. Of course, if one already subscribes to the tenets of cognitive semantics, meaning in an abstract domain like modality couldn't be anything other than a derivative, through metaphor or a similar process, of more basic concepts, and the fact that force dynamics provides a plausible metaphorical basis would be an argument in its favor. However, since we are evaluating and comparing different theories of modality, this is not a relevant argument for us.

In evaluating the force-dynamics view as a theory of modality, it is useful to separate deontic modality, as analyzed by Talmy, from epistemic modality, discussed by Sweetser. The analysis of deontic modality relies on the idea that there are psychosocial forces which we understand based on our understanding of physical forces. I find it quite plausible both the psychosocial forces really exist and that we have an understanding of them. (In fact, Winter and Gärdenfors 1995 and Gärdenfors 2007 are of the opinion that our concepts of such forces are basic and not derived from an understanding of physical forces.) Nevertheless, it is not sufficient to simply say that deontic modals express relations among forces in this domain. To see this, compare the following:

(152) (a) The wind pushed the boat across the lake.
     (b) The law compelled Julie to pay income tax.
     (c) The law compels Julie to pay income tax.
     (d) According to the law, Julie must pay income tax.

Example (152a) is a simple force-dynamical statement; (152b) is a close analogue in the deontic domain. The third example, (152c), is more complex than (152b), in that unlike (152b) it does not entail that Julie actually paid her taxes. This sentence could describe a situation in which she just decided not to pay (she will probably get into trouble as a result). An analysis of (152c) must involve both a semantics for (152b) and an account of the component of meaning which disallows the inference that she paid. The modal statement (152d) is like (152c) in this respect; it does not entail that Julie paid income tax. The cognitive semantics theory describes the pattern in (152d) as a metaphorical extension of the one in (152a).

The fact that (152d) does not entail that Mary paid her taxes is one of the core points which motivated prior theories of deontic modality. In particular, within modal logic $\Box p$ does not entail p because the accessibility relation is not reflexive, and a similar explanation was carried over into other theories based on possible worlds. More generally, one can say that deontic modality is not about the relation between rules and the events which actually transpire, but rather about the relation between rules and hypothetical events. The need to talk in a clear way about hypothetical events is precisely why possible worlds were introduced into semantics. As far as I can see, the force dynamic theory is unable to explain this fundamental property of modality.[34]

The force dynamical account of epistemic modality suffers from the same problem as that of deontic modality. For example, (151a) does not describe situations in which the weight of evidence forcing the speaker to believe that it's raining is greater than any opposing tendency not to believe. Rather, it describes a situation in which the weight of

---

[34] Langacker has noticed this issue, saying that in a modal sentence we are describing a situation in which some individual has a potency to perform some action or someone has an inclination to accept a belief. For example, in the case of epistemic modality, he says:

[The] profiled relationship has not been incorporated into the speaker's reality conception. It is however a candidate for acceptance. It is under consideration, and the speaker inclines towards accepting it with varying degrees of force, reflected in the different modal choices.... [This] requires mental effort and engenders a force dynamic experience.

(Langacker 2006: 21–2)

These unrealized potencies/inclinations are indicated in Langacker's graphical representations of meaning with dashed lines, rather than solid ones. However, Langacker terminology and notation do no more than identify the issue, and there is no analysis of it within the cognitive semantics framework.

evidence should, as an epistemic ideal, force that belief. Notice that my paraphrase of the meaning itself involves the expressions: *as an epistemic ideal* and *should*. These expressions point at what is essentially modal in the meaning of *must*, and the force dynamics metaphor fails to provide any analysis of them.

Another issue worth discussing in the case of epistemic modality concerns the nature of the antagonist's force, the "weight of evidence" pushing the agonist to adopt a certain belief. What sort of force does evidence apply towards a particular belief? In the case of the strongest and weakest modals like *must* and *may*, it seems that Kratzer's theory gets the relation right. If a body of evidence entails p, then the "force" to adopt p is sufficient to compel belief (as an epistemic ideal), and if a body of evidence is compatible with p, it does not prevent a belief that p (as an epistemic ideal). This point leads me to suspect that, if one were to work out Sweetser's ideas in detail, they would be reducible to Kratzer's theory. In particular, the relevant notion of force may be nothing but logical relations.[35] If this is right, the unique contribution of the cognitive semantics theory would be the claim that our understanding of these logical relations is derived from force dynamics through a metaphorical process.

It is worth pointing out that Sweetser has co-authored another work relevant to modality, Dancygier and Sweetser (2005). This book does not focus on modal expressions, but rather only mentions them in connection with an analysis of conditional constructions. Because it does not present a well-developed theory of modality, I will not discuss it further here, except to note that it is presented in terms of a "mental spaces" model.[36] Dancygier and Sweetser (2005) do not develop Sweetser's earlier proposals concerning the semantics of modality, and so it is not clear whether she continues to espouse the metaphor-based, force-dynamical analysis.

---

[35] Of course, the intermediate cases of graded modality (e.g., *probably*, *there is a slight possibility that*) are more difficult to analyze. As we saw, Kratzer offers an explicit theory of many such cases, and we might look to a theory based on probability for an analysis of these forms as well, e.g., Frank (1996); Swanson (2006a, 2007); Yalcin (2007a). These theories provide another attractive starting point for an attempt to render Sweetser's ideas more precise.

[36] See Fauconnier (1994) for more on mental spaces; the mental spaces are very similar to the discourse representations of DRT, cf. Kamp (1981) on DRT generally and, e.g., Asher (1986); Roberts (1987, 1989), on modality.

3.3.2 The cognitive-functional response to formal semantic analyses of modality

Although research on modality within cognitive linguistics has not engaged very much with prior or contemporary work based on possible worlds, the cognitive linguists have made a few criticisms of the standard approach. In many cases, the criticisms amount only to an argument that the paradigm of cognitive semantics is superior to the broader framework of formal semantics (of which possible worlds semantics is an important part). This point is made concisely by Dancygier and Sweetser (2005) in discussing Situation Semantics:

> But its basis in an objective truth-based semantics is strongly at variance with the claims of Mental Space Theory, which assumes that only experientially based construal—rather than objective truth—is accessible to human systems of meaning and interpretation.   (p. 11)

This criticism shows a deep misunderstanding of possible worlds semantics, however. The term "objective truth" suggests two related concerns. I will briefly discuss each in turn.

The first issue concerns truth. Cognitive semanticists assume that formal semantics explains meaning in terms of whether sentences are in fact true or false, but this is incorrect. Formal semantics always defines its semantic terms, like truth, relative to something, for example a possible world, a situation, a frame, or a model. That is, it is concerned with "truth in a world" or "truth in a model," not simply "truth." In possible worlds semantics, the meaning of a sentence can be identified with its character (the function from contexts to propositions), with the proposition it expresses (that is, the set of worlds in which it is true), or the related dynamic concepts like update potential. The proposition or update potential which serves as the meaning of a sentence represents informational content. As we noted in the discussion of dynamic semantics in Section 3.2, this content can be viewed either psychologically or pragmatically. According to the psychological view, the possible worlds are used to represent the content of an individual's beliefs or knowledge; on the pragmatic view, they are used to represent the pragmatic presuppositions of the conversation. In either case, whether a sentence is in fact true is beside the point.

The second objection concerns objectivity. Cognitive semanticists are critical of formal semantics for allegedly assuming that the bases for assigning sentences to one of the categories "true" or "false" are facts about the world which are so independently of humans' concepualization. For example, if we say that *it is raining* is true in a given world *w*, this is seen as involving an assumption that the existence of falling rain is an objective fact about the world, rather than a structure projected onto the world in some way by the human mind. In fact, however, formal semantics is neutral on the question of objectivity in this sense.[37] The possible worlds that figure in our theory need not be the set of metaphysically possible worlds (if it even makes sense to talk about such a set). As far as formal semantics goes, possible worlds could be defined in terms that take into account psychological and social factors. A relevant comment is made by Barbara Partee, a prominent semanticist in the formal semantics tradition:

> A non-absolutist picture seems to fit linguistic semantics better than an absolutist one, where by absolutist I mean the position that there is one single maximal set (or class, if it's too big to be a set) of possible worlds. If a philosopher could find arguments that in the best metaphysical theory there is indeed a maximal set, I suspect that would for the linguist be further confirmation that his enterprise is not metaphysics, and I would doubt that such a maximal set would ever figure in a natural language semantics. (Partee 1988: 118)

There are a number of other criticisms of theories of modality based on possible worlds which are more specific to the topic of modality. I will briefly discuss four. I derive these criticisms from the the writings of Sweetser (1990) and Coates (1983), but I believe their evaluations to be representative of the functional/cognitive literature in general.

First, Sweetser (1990) states that the approach of Kratzer (1977) and similar work "has been essentially to subsume the root meaning of the modals under very general epistemic readings; thus root 'can' comes to refer to logical compatibility between a person's (or the world's) state and some event, while root 'must' refers to the logical necessity of the occurrence of some event, given the state of the world" (p. 56). In light of the discussion of Kratzer's theory in Section 3.1 it's easy to see why this criticism misses the mark. To begin with, while it is true that on Kratzer's view the meanings of all modals involve logical relations, states and events are not the sorts of things which figure in these relations. Rather, these relations hold between propositions and conversational backgrounds. The conversational backgrounds are all formally the same kind of thing (functions from worlds to sets

[37] I for one have no problem with the assumption that whether it's raining or not is an objective matter, and if some theory implied that it's not, I would take that as a strike against the theory. But my point here is that formal semantics does not itself commit one to objectivity.

of propositions), but they are not all the same. For example, while a dynamic modal utilizes a circumstantial background as its modal base, an epistemic modal utilizes an epistemic one. There is no sense in which the dynamic case is reduced to the epistemic one. Rather, they are given a unified analysis in terms of conversational backgrounds.

With the above corrections, Sweetser's descriptions of Kratzer's analyses of dynamic *can* and *must* come close to correctly characterizing the early sections of Kratzer (1977). (More accurately, modality was defined there in terms of compatibility or entailment between the set of propositions determined by a conversational background, at a given world, and a proposition. *Can* expresses logical compatibility, and *must* logical entailment, between the set of propositions and the proposition.) However, it is not an accurate description of Kratzer's theory overall (e.g., the later sections of Kratzer 1977, Kratzer 1981, and Kratzer 1991b). As we have seen, in her later work the theory involves a wider range of modal relations—necessity, weak necessity, possibility, slight possibility, and so forth—which are more subtle and crafted to serve the needs of a theory of modality. It also adds the parameter of the ordering source, so that each use of a modal is sensitive to multiple conversational backgrounds at once.

At the root of Sweetser's objection is probably a belief that logical notions are essentially epistemic in nature. This is a misconception, but one which could easily arise from the fact that these notions, as employed in traditional logic, are indeed important because of their role in developing a logical system, that is a formal theory of valid inference. But formal semantics does not use these logical concepts for the same purposes as logic itself, a point stressed above in Chapter 2. Rather, it uses them as tools for creating a semantic theory. The fact that these tools have their origins in logic doesn't turn any theory which uses them into a form of logic, any more than my use of a hammer to crack nuts turns me into a carpenter.

Sweetser makes a second, more significant criticism of theories of modality based on possible worlds. She claims that, because they treat deontic and epistemic meaning as based on the same relations involving possible worlds, such theories cannot explain the historical tendency for epistemic meanings to derive from deontic ones.[38] In contrast, the

---

[38] It has been claimed that there are exceptions to this tendency (Narrog 2005; Traugott 2006). We should be careful about evaluating theories against this tendency until its exact parameters become clear.

cognitive theory argues that this tendency follows from the fact that deontic meanings are relatively closer to fundamental force-dynamical concepts; in other words, a metaphorical process can take us from force dynamics in the physical world to concepts of obligation and permission, and a subsequent process can take us from there on to epistemic necessity and possibility, but there's no direct link from forces acting upon objects to these epistemic meanings. (I'm not sure why there should be no direct link of the latter sort, but we can take it as a stipulation.)

The metaphor-based theory is not the only one which aims to explain the pattern of historical change with modals. For example, Traugott (1989, 2006) and Traugott and Dasher (2002) argue that the epistemic meaning of *must* derived from the deontic one because of pragmatic processes operating in contexts in which both readings were plausible. Traugott and Dasher (2002) give the following example (their (43), p. 127, cited from Warner 1993):

(153)   Ealee we moton sweltan.
       all   we must   die
       'We all must die.'

Paraphrasing Warner's description, the context of this sentence provides a meaning like "If the Jews do not leave Egypt, we (Egyptians) all must die." Here the phrase *we all must die* is ambiguous: the deontic meaning is "we are all obliged to die by God's will," and the epistemic meaning is "we all certainly will die." The deontic reading may be difficult to perceive in the modern English translation but, as Traugott and Dasher emphasize, this should not blind us to the fact that it was possible in earlier stages of the language.

According to Traugott and Dasher, the deontic and epistemic interpretations of (153) are linked by inference: if we are all obliged to die by God's will (deontic), we know that we all will die (epistemic). In a nutshell, Traugott and Dasher's claim is that the epistemic meaning for *must* arose over time because inferences of this kind became absorbed into the word's core semantics. Such a claim fits into their broader picture of how semantic change works and, while they do not discount the relevance of metaphor to semantic change in general, they provide arguments against the metaphor theory and in favor of the inference-based one in the case of English modals.

TABLE 3.4. Readings of (153) in Kratzer's semantics

Modal base	Ordering Source	Paraphrase
Circumstantial	Deontic (God's will)	"We are obliged to die."
Circumstantial	Empty	"We have no way to avoid dying."
Circumstantial	Stereotypical	"We are definitely going to die."
Epistemic	Empty	"We are certain to die."
Epistemic	Doxastic	"We very probably will die."

The issue of whether the correct theory of the semantic change of modals is inference-based or metaphor-based, or some combination of the two, is beyond the scope of this book (and beyond its author's competence). The key point here is only that there is an alternative theory of semantic change in modals which is perfectly compatible with semantics theories of modality based on possible worlds. In terms of Kratzer's (1977) theory, we can describe (153) as a case in which the context is compatible with either a deontic or an epistemic conversational background; in terms of the more sophisticated Kratzer (1981), this sentence could have either a circumstantial modal base and deontic (or empty, or stereotypical) ordering source, or an epistemic modal base and an empty (or doxastic) ordering source. Table 3.4 provides paraphrases of the various combinations.

The kind of semantic theory of modals offered by Kratzer, employing an underspecfied core meaning and multiple sources of contextual information, seems well matched with Traugott and Dasher's proposal. At the earlier stage, the lexical meaning of the modal might be compatible only with a circumstantial modal base. But contexts like (153) would be ones in which the addressee inferred another meaning, an epistemic one. Moreover, sometimes the speaker might have intended this additional inferred reading, and might have expected the hearer to recognize this intention. In Kratzer's theory, the non-epistemic and epistemic meanings differ only in the choice of conversational backgrounds. That is, if (153)'s modal base had been epistemic, the sentence's meaning would have been equivalent to the inferred reading. Over time, the lexical restriction against *must* being interpreted with an epistemic modal base went away. This change served to bring the semantics of sentences involving *must* more in line with the meanings inferred by hearers in cases like this one.

Next we turn to the criticisms of previous approaches to modality made by Coates (1983). While her points are not specifically directed at accounts based on possible worlds, it appears likely that she intends for them to apply to such accounts. She summarizes the two issues she wishes to raise as follows:

[F]irst, the terms and categories traditionally used (e.g. aletheutic, epistemic, deontic) are opaque and intimidating; secondly, modal logic emphasizes objective modality at the expense of subjective modality.   (p. 54)

Coates's first objection against theories of modality based on possible worlds is that the terms and categories it employs are opaque and intimidating. The first point I would like to make here is that, since one of the main goals of this book is to present in a clear but detailed way the major theories of modality, I hope very much that no one reading at this point finds the terms and categories intimidating or opaque (any more). But in any case, though I see how they may have been intimidating, I don't at all see the point that they are opaque. The modern descriptive literature on modality is quite clear, I believe, for example the work of Lyons (1977), Perkins (1983), and Palmer (1986, 1990, 2001), and the use of the same concepts in modal logic is a paradigm of clarity, although one must learn sufficient modal logic before this becomes apparent. And, second, though I'm not sure how seriously to take Coates's argument that terms and categories should be discarded because they are intimidating, it is worth insisting that the difficult and perhaps intimidating nature of modal logic is no reason for serious scholars to turn away. After all, no one would object to using advanced mathematical concepts in the natural sciences on the grounds that they can be intimidating.

Coates's other objection is more important. A number of functional linguists have argued that modals expressions, in particular epistemic modals, have a subjective or intersubjective meaning more or less in the sense introduced above in the discussion of Nuyts (2001b). Coates points out that the concept of subjectivity has not played a role in modal logic, and suggests that semantic analyses in general cannot handle subjectivity. Therefore in the next section let us look at the various notions of subjectivity which have been discussed in the functional linguistics literature. Once we have seen why a notion of subjectivity is important to the theory of modality, we will then have to consider whether subjectivity argues in favor of a functional/cognitive semantics, or whether it can be incorporated fruitfully into formal theories based on possible worlds. As we will see in the next section, and in more detail in Section 4.2.1, I believe that the latter is correct.

3.3.3 Subjectivity and intersubjectivity

Lyons (1977) uses the following example to explain what he means by subjectivity in epistemic modality (his (14), p. 797):

(154)   Alfred may be unmarried.

Lyons distinguishes two different ways in which a speaker can use (154). In most situations, Lyons says, "the speaker may be understood as subjectively qualifying his commitment to the possibility of Alfred's being unmarried" (p. 797). However, Lyons suggests that we also consider a situation in which Alfred is a member of a community of 90 people, and we know for certain that this community contains 30 married people. In this case, Lyons suggests, it is possible to use (154) to state that there is an objective probability (specifically, 2/3) that Alfred is unmarried. The following examples might make it easier to see Lyons's intuition:

(155)   (a)   Alfred has smoked for 30 years, so I worry about him. He may well get lung cancer.
        (b)   Alfred has smoked for 30 years, and the statistics tell a scary tale. He may well get lung cancer.

(155a) has a subjective use, while (155b) is objective. In the former, *he may well get lung cancer* seems to mean "the possibility of his getting lung cancer is one to worry about," while in the latter it has a meaning more like "the probability of his getting lung cancer is relatively high."

Lyons (1977) thinks that the concept of subjectivity applies to deontic modality as well, but he does not have an extensive discussion of this case. Palmer (1986, 1990) develops the idea that a deontic modal can be subjective or objective. Consider (156). In the subjective (156a), the speaker is the source of the obligation to leave, while in the objective (156b), the obligation to be charitable is due to moral principles which do not depend on the speaker:

(156)   (a)   You must leave now (because I say so).
        (b)   The rich must give to the poor.

Later authors working in Lyons's tradition have expressed doubt that epistemic modality is ever objective. For example, Palmer (1986) notes that if we define epistemic modality in terms of the speaker's (or speaker's and hearer's) knowledge or strength of commitment, and subjectivity in terms of the evaluation of a proposition by the speaker (or speaker and hearer), epistemic modality will always be subjective in this sense. The closest thing to objective epistemic modality on Palmer's view would be alethic modality. Lyons's view, too, seems to be that alethic modality in natural language is simply an extremely objective version of epistemic modality.

In any case, there is a meaning difference of some sort apparent in (154) and (155). I will retain Lyons's labels of "subjective" and "objective" for the time being, and turn to the question: what is the nature of the distinction between subjective and objective epistemic modality? Lyons believes that the two are fundamentally different in both their semantics and pragmatics. In particular, he believes that subjective epistemic modality operates at the level of speech acts (see also Palmer 1990). To see how this works, let's first examine Lyons's analysis of a non-modalized assertion:[39]

(157) (a) Alfred is unmarried.
       (b) I-SAY-SO + IT-IS-SO + [[Alfred is unmarried]]

The speech act of assertion is analyzed into three parts, as in (157b). When we have an epistemic modal, one of the first two parts of will be replaced by a modal concept. A subjective epistemic modal will lead to a change in the first part, whereas an objective epistemic modal will lead to a change in the second part. Here are Lyons's representations for (154):

(158) (a) **Subjective:** Poss + IT-IS-SO + [[Alfred is unmarried]]
       (b) **Objective:** I-SAY-SO + Poss + [[Alfred is unmarried]]

According to Lyons, a speaker who uses an epistemic modal subjectively is no longer categorically "saying-so," since POSS has replaced I-SAY-SO. This is meant to represent the idea that the speaker is not fully asserting that Alfred is unmarried, but rather expressing some reservations. In contrast, a speaker who is using an epistemic modal objectively is saying that something is definitely so, just like in a regular assertion. Though this systemization of the two meanings is extremely interesting, two points remain unclear: First, do the Poss in (158a) and the Poss in (158b) have anything in common? And, second, what is the nature of the speech act is in (158a)? If the speaker is not asserting, what exactly is she doing?

Kratzer (1981) discusses the distinction between subjective and objective modality as it is described by Lyons, but analyzes it in purely

---

[39] Lyons writes (157)b as ... p, but I will use a more transparent notation.

semantic terms. Notions of speech acts and assertion are not relevant. She discusses the following pair of German examples (her (42)–(43)), in the scenario that "Lenz, who often has bad luck, is going to leave the Old World by boat, today, on Friday thirteenth" (p. 57):

(159) (a) Wahrscheinlich sinkt das Schiff
            Probably   sinks the boat.
            'Probably, the boat will sink.'
       (b) Es ist wahrscheinlich, daß das Schiff sinkt.
            It is probable   that the boat sinks.
            'It is probable that the boat will sink.'

Someone saying (159a) may base her statement on the idea that Friday the thirteenth is a day of bad luck. Even if she admits that this belief is a mere superstition, she can still use it as a grounds for asserting (159a). In contrast, this person could not say (159b) on these grounds. In order to sincerely assert (159b), she must have in mind only reasons having to do with the weather, the condition of the boat, the likelihood of icebergs, and so forth—in other words, only reasons which are held in high regard within educated society. In terms of Kratzer's theory, this difference can be analyzed by saying that *wahrscheinlich* allows a stereotypical ordering source which includes superstitions; such propositions cannot be defended in educated society, and we can call the resulting ordering source "subjective." In contrast, *es ist wahrscheinlich* only allows propositions defensible in educated society in its ordering source. Such propositions comprise an "objective" ordering source.

Kratzer's discussion is probably not adequate as a general theory of the difference between subjective and objective epistemic modality. She describes the difference between subjective and objective ordering sources in a highly culture-specific way, since it depends on the status of certain beliefs as superstitions. Perhaps in another culture, the kinds of beliefs that we consider to be superstitions are esteemed as the most indisputable kind of knowledge. What we need is a general theory which lets us classify some reasons as objective and others as subjective on the basis of how those reasons are viewed, employed, and responded to by speakers of the language in question. For example, the fact that among German and English speakers one can dismissively say "that's a mere superstition" shows that fear of Friday the thirteenth is subjective, and the fact that one cannot dismissively say "that's a mere scientific fact" shows that a satellite image of a giant iceberg is objective. But in another language community, the relevant distinction between kinds of reasons might be quite different.

While Lyons and Kratzer would probably more or less agree on which uses of epistemic modals should be counted as objective or subjective, their analyses are quite different. Lyons's proposes that subjective epistemic modality affects the nature of the speech act, and not the proposition expressed. Kratzer's analysis, in contrast, treats epistemic modality as making a semantic contribution to the proposition expressed. Recent work in the formal semantics tradition has returned to this issue. For example, Papafragou (2006) supports Kratzer's position, while von Fintel (2003), von Fintel and Gillies (2007b), and Portner (2007a) suggest that epistemic modals both contribute to semantics and have an effect at the speech act level. When we discuss epistemic modality in more detail in Chapter 4, we will return to this debate.

Research on modality within functional linguistics and cognitive semantics has been very interested in the notions of subjectivity and intersubjectivity; for example, see Coates (1983), Langacker (1985, 1990), Traugott (1989, 2006), Traugott and König (1991), Traugott and Dasher (2002), Nuyts (2001a,b), and Narrog (2005). Unfortunately, none of these authors seem aware that Kratzer's (1981) theory has been applied to subjectivity. For this reason, an argument along the lines of that presented by Coates (1983) that possible worlds semantics is unable to deal with subjectivity has never been mounted in a serious way. Moreover, as noted above, the phenomenon of subjectivity in Lyons's sense has been the subject of much recent research within possible worlds semantics. Therefore, the issue of subjectivity does not provide any easy argument in favor of cognitive/functional approaches.

There are two main approaches to subjectivity within the cognitive and functional literature:

1. The ideas of Traugott and her associates, based on a traditional understanding of the roles of semantics and pragmatics (though with an emphasis on the pragmatics).
2. The ideas of Langacker, developed within cognitive semantics and not involving a traditional conception of pragmatics.

Besides these two, competing approaches, there is a fair amount of work within functional linguistics which is based on a conception of subjectivity similar to Lyons's (for example, Coates 1983; Narrog 2005), or on a conception similar to Lyons's but augmented by ideas from pragmatics and sociolinguistics (for example, Nuyts 2001b). In addition, some functional work on modality defines epistemic modality in the same way as Palmer (1986), that is, in a way more or less equivalent to Lyons's definition of subjective epistemic modality (for example, Halliday 1970; Bybee and Fleischman 1995b).

Traugott's and Langacker's approaches are developed within quite distinct frameworks, but the most fundamental difference between them has to do with the status of pragmatics. Traugott analyzes subjectivity as the contribution of the speaker to meaning via pragmatics. For example, indexicals like *I* and *here* involve subjectivity because, through them, the speaker and the speaker's perspective play a role in determining meaning. In particular, they contribute to the proposition expressed: if someone says "I am happy," the proposition expressed depends on the identity of the speaker. Subjectivity can also be more purely pragmatic in nature, for example, in the case of performatives: if someone says "I order you to leave," the role of the speaker in creating meaning is clear. Traugott is also able to develop a healthy notion of intersubjectivity. A sentence is intersubjective to the extent that its meaning depends on the relationship between the speaker and addressee. A clear example is the contribution to meaning (pragmatic meaning, not semantic meaning in this case) of markers of politeness; moreover, it seems to me that performatives would also count as heavily intersubjective.

When applied to modality, Traugott's notion of subjectivity is very similar to those of Lyons and Palmer. In Traugott's view, modality is a marker of (inter)subjectivity to the extent that, in a particular case, it involves the speaker or speaker/hearer in determining semantic or pragmatic meaning. For example, epistemic modality is heavily subjective because it has to do with the speaker's knowledge. As far as I can see, Traugott's view is neutral between Lyons's idea that subjectively interpreted modals operate at the speech act level and Kratzer's analysis in terms of the proposition expressed. If Lyons is right, subjective modals are like performatives, while if Kratzer is right, they are like indexicals. In either case, they count as subjective for Traugott. (If Lyons is right, they should probably also be classified as heavily intersubjective.)

Langacker's theory is developed within cognitive semantics, and in tune with the emphasis of this approach, Langacker defines the concept of subjectivity in terms of the speaker's conceptualization of a situation. Because the theory does not give any formal role to the interaction between the speaker and the addressee, there is no pragmatics in the

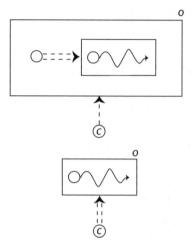

FIGURE 3.6. Langacker's representations of a desire verb (top) and an epistemic modal (bottom)

traditional sense.[40] Hence Traugott's ideas cannot be translated into cognitive semantics (see also Brisard 2006).

Langacker's concept about subjectivity begins with the idea that semantic representations may include the representation of a "conceptualizer" or "subject of conception," the individual whose cognitive state is being represented. Suppose that some diagram R provides a theoretical model of my mental representation of some situation. In some cases, R may contain a component c which represents me in my role as the one who hosts this representation in my mind. This c is the conceptualizer. The rest of R apart from c is the object of conceptualization, o. In the kind of metaphorical terms that cognitive semanticists use to explain this concept, we can say that c provides the perspective on a situation o. Figure 3.6 (based on Langacker 2006) presents simplified versions of the representations associated with a desire verb and an epistemic modal.

Let us apply the upper diagram in Figure 3.6 to (160a): c is conceptualizing a situation o in which John wants to leave. Within o, John (the little circle) experiences a metaphorical force, indicated by the double-lined arrow, tending to cause him to leave. The fact that this force is only potential, and may not make him leave, is indicated by the fact that the double-lined arrow has dashed and not solid lines, and the inner box represents the hypothetical situation in which he leaves.

---

[40] Verhagen (2005, 2006) has attempted to incorporate the addressee into the cognitive semantics framework. Space precludes me from discussing his ideas.

(160) (a) John wants to leave.
       (b) John must be leaving.

The epistemic sentence (160b) is represented by the diagram on the bottom of Figure 3.6. Here, c conceptualizes o, a situation in which John leaves. The metaphorical potential force in this case is derived from c's knowledge, and tends to cause c to accept o as part of his knowledge. Thus, the double-lined arrow indicating the force is not within o, but between c and o.

The conceptualizer stands in the first instances for the speaker, but it also stands for the addressee in the sense that the addressee will be able conceptualize things in the same way as the speaker once he or she understands what the speaker is saying. (This role for the addressee does not amount to introducing real pragmatics into the theory, however, since there is no cognitive representation of interaction between speaker and addressee. The speaker and addressee are master and apprentice conceptualizers, one might say, with the speaker showing the addressee how to conceptualize the situation.)

As modal meanings changed over history from ones which would be represented as in the top diagram of Figure 3.6 and ones which would be represented as in the bottom diagram, two things happened according to this approach: there is less material in o and some of what was previously in o, the force dynamics, is now in the relationship between c and o. Both of these changes contribute to subjectification, in Langacker's sense, because they make the subjective construal of o by c relatively more important in the overall representation.

Now that I have sketched how Langacker analyzes the notion of subjectivity, let us think about what it has to offer. On the positive side, by introducing the concept of a conceptualizer it offers a structural parallel among pre-modal meanings, like those of desire verbs, and modal meanings which is more precise than that in Talmy's and Sweetser's version of the theory. On the negative, it inherits all of the difficulties of their theories, in particular the inability to analyze (as opposed to merely label through the use of dashed lines) the fact that the metaphorical force need not have its effect in the real world and the lack of clarity concerning the nature of the force, particularly in epistemic cases.

Langacker's proposal has an additional problem as well. What does it mean for the conceptualizer to be part of the representation of what the conceptualizer means? To put this point concretely, consider the representation at the bottom of Figure 3.6. The box o is the meaning of the material under the scope of the modal, and the whole diagram with c, o, and the arrow between them is the meaning of the modal sentence. The essence of the semantics of the sentence is the relationship between c and o indicated by the arrow. But don't forget that this representation is supposed to be in the speaker's mind. Therefore, the relationship between the speaker and o consists in o being a part of this diagram which is in the speaker's mind. At this point, I don't see how to make sense of the representation any more. The two relationships which the speaker is supposed to have to o are incompatible. Metaphorically speaking, the relationship between the speaker and o is not "inside the page" along the axis of the arrow, but rather one which allows her to "see" o from an angle which also takes in c and the arrow.

This problem arises because cognitive semantics does not make a distinction between the cognitive representation of a meaning and the meaning itself. There would be no problem with saying I am in one relationship with the situation/event/proposition represented by o (this would be the relationship indicated by the dotted arrow in the diagram) and another with the representation itself (this relationship would consist in the representation being in my mind). But cognitive semantics cannot think this way, since the representation and the meaning are identical.[41]

3.3.4 Evaluation of ideas about modality in cognitive and functional linguistics

The force-dynamics theory of modality developed by Talmy, Sweetser, and Fauconnier contains an attractive intuition for how modal concepts relate to more basic, non-modal concepts. However, it does not provide an analysis of two central features of modal meaning: the fact

[41] My wife and I found our three-year-old son's fish on the floor. It had jumped out of the tank, and of course we discarded the fish so he wouldn't see it. We couldn't tell him what had happened, just saying "Maybe he jumped out." But he didn't want to believe the worst had happened. Maybe the fish was just hiding. He asked "Did you see it on the floor?" We lied, "No." "Then he probably didn't die." I don't think this little boy (smart though he is) was able to conceptualize himself as having strong evidence for believing that the fish was alive. He was only evaluating the evidence, saying that the fact that we didn't see the fish on the floor is a good reason for believing that the fish is alive. He was not describing himself as someone who has a good reason for believing that the fish is alive.

that the meanings make reference to hypothetical situations, and the nature of "forces" of authority, desire, evidence, and the like which relate to those hypothetical situations. For this reason, it remains unclear how to derive detailed linguistic consequences from this theory.

Let me give an example of this point. It seems contradictory for someone to assert the following:

(161)   #It is raining and it must not be raining.

The status (161) can be easily explained within any version of possible worlds semantics, but I do not see how to derive, within the force-dynamics theory, this or other concrete facts about the use of specific modal sentences. In particular, why is this sentence contradictory under normal circumstances? To answer this question, the cognitive semanticist will need to say something about what kind of metaphorical forces can be present when one's mind contains a representation for *It is raining*. One possible story looks like this: the mind is so constructed that, when it contains a fully integrated and believed representation of the meaning of *It is raining*, it cannot also contain a metaphorical cognitive force pushing the individual to believe that it is not raining (or, at least, no such force strong enough to meet the threshold for using *must*). But why not? There are really two problems here. First of all, this is just one case, and what we need is a general understanding of how metaphorical cognitive forces arise from beliefs (and desires, dealing with people in authority, and so forth). And, second, it's not really plausible that someone who believes that it's raining can't feel a metaphorical force pushing them to believe that it's not. Certainly it could happen—even priests have doubts, they say. But even if your mind is in this unfortunately confused state about whether it is raining, you know better than to say (161) to someone. The real problem is that (161) is contradictory, and the rules of conversation do not approve of such blatant inconsistency.

Cognitive and functional approaches to modality make an important contribution with their focus on the notions of subjectivity and inter-subjectivity. While they share an interest in this topic with descriptive linguists like Lyons and Palmer, they develop it in a number of quite different ways. Lyons sees subjectivity as related to speech-act theory. In contrast, most functional linguists relate it to pragmatics and socio-linguistics (e.g., Coates, Nuyts, Traugott), while cognitive linguists relate it to the conceptualization of a situation. The key difference between these views has to do with the status of pragmatics. The functional approach develops the traditional perspective of pragmatics as focusing on meaning which arises from the interaction among speakers, addressees, and the context. Subjectivity is the relevance of the speaker to meaning, and intersubjectivity is the relevance of the speaker-addressee relationship to meaning. The cognitive approach does not appeal to traditional pragmatics in this way, and indeed would have a difficult time doing so given that the meaning of a sentence is mental construct, and so is something that can exist in a single individual's mind. Instead, subjectivity is defined in terms of the status of a conceptualizer (in the first instance the speaker, in a secondary way, the addressee) within the semantic representation. This attempt to put the speaker into the semantics causes just as many problems as the need, discussed in the previous paragraph, to explain the metaphorical forces relevant to modality in cognitive terms. In both cases, the fundamental problem for the cognitive semantics theory is a lack of a real pragmatics, and thus ultimately a failure to distinguish semantics from pragmatics.[42]

3.4 Looking ahead

In this chapter, we have surveyed three major theories of modal semantics. Let me summarize some of the main conclusions which I believe we should draw:

1. The possible worlds semantics for modality, originally developed within modal logic, provides the best theoretical basis for understanding the fundmental nature of modality.
2. Kratzer's theory develops the ideas of modal logic in ways which allow for a detailed analysis of different kinds of modal meanings in natural language.
3. Dynamic logic, though it is restricted to a narrow range of phenomena, offers important tools for understanding the meanings of modals in discourse.

---

[42] Let me emphasize that this point is independent of any particular ideas about the relationship between semantics and pragmatics. I am emphatically not suggesting a simple "semantics first, pragmatics second" model. What's needed is just some notion of semantic content(s), and a theory of how those contents are put to use which involves (perhaps among other things) the speaker's intentions toward, and interactions with, the addressee.

4. Descriptive, functional, and cognitive approaches to linguistics do not provide a convincing theory of modal semantics. However, they investigate in detail the important concepts of subjectivity and intersubjectivity.

In this chapter, the goal has been to gain a understanding of both the ideas and techniques of several important theories of modality. In the next two chapters, we'll focus instead on issues in the semantics of modality, using what we have learned about the theories to support our thinking about these issues.

4

Sentential Modality

In Chapter 1, I distinguished three linguistic domains in which modality operates: sentential, sub-sentential, and discourse. Sentential modality operates, as the name suggests, at the level of the entire clause. Most treatments of modality take sentential modality to be the central case; for example, in English, we see an emphasis on the semantics of modal auxiliaries. While these modals may be important at the sub-sentential and discourse levels as well, they clearly are grammatically linked to the sentence, and they have an obvious effect on sentence meaning. They interact syntactically and semantically with other clause-level elements, such as negation, tense, and aspect. It is their sentence-level meaning and sentence-level grammar which we aim to understand better in this chapter.

Syntactic analyses of sentential modality show that it is one of the meaningful grammatical elements which associates with the predicate-argument core of the sentence. That is, we typically think of the core of a simple sentence like (162) as the verb *arrive* and its subject *Mary*:

(162)   Mary arrived.

Even within this very simple English sentence, we do not have the core alone. We also see the past tense, a meaningful functional element, combining with this core. In a more complex example like (163), there are additional meaningful functional elements, *should* and *not*:

(163)   Mary should not arrive.

Should and *not* combine with the core to produce (163); somehow, in accord with the principles of English syntax, *Mary* ends up as the first word.

There are various ways to represent these non-controversial ideas about the sentence-level syntax of modals, and so for concreteness I will make a few assumptions: First, that the core of the sentence is a projection of the verb, VP. Second, that each meaningful functional element F projects its own full phrase, FP. For example, the modal M projects a modal phrase MP. And, third, that the subject reaches its position at the beginning of the sentence via movement, leaving behind a coindexed trace. Thus, (163) can be represented as in (164):

(164)

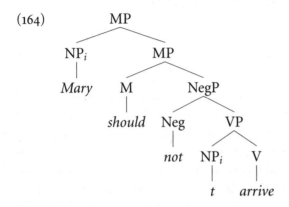

The fact that I use recursion of the MP to host the subject is not important; there may be other functional projections in some or all modal sentences, for example tense, aspect, or voice, and it's a difficult question where exactly the subject is.

In this chapter, we will focus on the semantic classification of modal expressions (Section 4.1) and the semantics of the various classes (Sections 4.2–4.4). We have already touched on much of the relevant material in Chapters 2–3, but our emphases there and here are different. In Chapters 2–3, our goal was to learn about several important theories of modality, and language data was introduced in a way designed to serve that end. Here and in the next chapter, our goal is to understand what these theories have to teach us about language.

4.1 Semantic categories of sentential modality

The first topic we will discuss concerns systematic differences among semantically coherent groups of modals. We have already touched upon some important groups, for example epistemic and deontic modals, but we have not considered in an explicit way precisely what classification or classifications are appropriate. The goal of this section is to consider the alternatives and settle on the best method of classification.

4.1.1 Epistemic, priority, and dynamic modals

I propose three primary categories of sentential modality: epistemic, priority, and dynamic. Epistemic modals are those pertaining to the speaker's knowledge, for example (165).[1] In terms of Kratzer's theory, epistemic modals have an epistemic modal base and various kinds of ordering sources (doxastic and stereotypical ones are common).[2] Priority modals include the deontic, bouletic, and teleological (goal-oriented) modals, as in (166). The idea behind the term "priority" is that such things as rules, desires, and goals all serve to identify some possibility as better than, or as having higher priority than, others. Priority modals have a circumstantial modal base and fairly easy-to-perceive ordering sources which provide the priority ranking. In the examples in (166), we have deontic, bouletic, or teleological ones. Dynamic modals also involve circumstantial modal bases, and fall into two primary sub-groups: volitional and quantificational. The volitional modals have a range of meanings (ability, opportunity, disposition) which have to do with the ways in which circumstances affect the actions available to a volitional individual, while quantificational modals seem to produce existential or universal quantification over individuals.

(165) **Epistemic**

    (a)   A typhoon may hit the island.

    (b)   Mary must have a good reason for being late.

(166) **Priority**

    (a)   The rich must give money to the poor. (deontic)

    (b)   You should try this chocolate. (bouletic)

    (c)   You could add some more salt to the soup. (teleological)

(167) **Dynamic**

    (a)   Volitional

          (i)   John can swim. (ability)

          (ii)   You can see the ocean from here. (opportunity)

          (iii)   Mary will laugh if you tell her that. (dispositional)

---

[1] This classification is first discussed in Portner (2007b).

[2] If natural language sentential modals can express alethic or metaphysical modality, these uses fall into the same class as epistemic modality. In that case a term like "factual modality" could be used to cover the epistemic, alethic, and metaphysical subtypes.

(b) Quantificational

(i) A spider can be dangerous. (existential)

(ii) A spider will be dangerous. (universal)

We have not looked in detail at quantificational modals before, so let's consider (167b). Both of these examples have an indefinite noun phrase as the subject, but this subject seems to receive different interpretations in the two examples: (i) has a meaning similar to *Some spiders are dangerous*, while (ii) has a meaning closer to *All spiders are dangerous*. The difference between "some spiders" and "all spiders" here correlates with the choice of modal, and is reminiscent of other cases in which the quantificational force of an indefinite seems to come from elsewhere in the sentence, such as the adverbs in (168):

(168) (a) A spider is *sometimes* dangerous.

≈ "Some spiders are dangerous."

(b) A spider is *usually* dangerous.

≈ "Most spiders are dangerous."

(c) A spider is *always* dangerous.

≈ "All spiders are dangerous."

The first analysis of indefinites and adverbs which could explain this pattern was given by Lewis (1975). (See also Kamp 1981; Heim 1982; Berman 1987; Groenendijk and Stokhof 1990; Heim 1990; de Swart 1993; Chierchia 1995a; von Fintel 1995, among many others.) Lewis proposed that the adverb quantifies over individuals of the kind described by the indefinite—in the case of (168c), for example, *always* provides universal quantification over spiders, yielding the meaning "all spiders are dangerous." In light of this analysis, he dubbed the relevant class of adverbs ADVERBS OF QUANTIFICATION. While some subsequent work (to be discussed below) has rejected Lewis's theory, his central observations— that adverbs of quantification are quantifiers, and that indefinites have a chameleon-like character when combined with them—have remained.

Given the similarity between (167b) and (168), Carlson (1977), Heim (1982), and Brennan (1993) have proposed that some modals function similarly to adverbs of quantification. Such modals can be called "quantificational modals."[3] I agree with this idea, though not with the specific

---

[3] Of course, according to possible worlds semantics, other modals quantify over possible worlds, and so all modals are properly described as quantificational. I use the term "quantificational modal" to describe those modals which appear to quantify over individuals, e.g., spiders in the case of (167b).

way in which they work it out. In particular, I think that quantificational modals and adverbs of quantification quantify over situations— not worlds (as other modals do) or individuals (as Carlson, Heim and Brennan thought).

Subjectivity and performativity

There are two other distinctions which will be important as we seek to better understand the various categories of modality. One is between subjective and intersubjective (or objective) uses of epistemic modals. These concepts were discussed in Section 3.3, and will come up again in Section 4.2. As mentioned in Section 3.3.3, some linguists apply the concept of subjectivity to deontic modals or even treat it as definitional of modality in general. While there is something to this perspective, I believe that it is counterproductive to call every kind of speaker- or addressee-orientation "subjectivity." Overuse of the term makes it difficult to talk about the particular contrast within the category of epistemic modality between cases in which the source of information is private and ones in which it is shared or of a kind that is considered highly reliable. It is this range of distinctions (which remains to be understood better) which I will refer to by the terms "subjective," "intersubjective," and "objective." Other notions of subjectivity will have to get other names.

 I refer to a modal as PERFORMATIVE if, by virtue of its conventional meaning, it causes the utterance of a declarative sentence to perform a speech act in addition to, or instead or, the act of assertion which is normally associated with declarative clauses. For example, Ninan (2005) argues that a sentence with deontic *must* (when used literally in a root declarative clause) both asserts something and performs a directive speech act. (Palmer 1986 labels a deontic modal of this kind "subjective," but, as mentioned above, I prefer to use the term in a narrower sense.) For example, according to Ninan, (169) performs a speech act similar to the imperative *Leave!*

(169)   You must leave.

Crucially, this imperative-like speech act is said to be part of the conventional meaning of *must*, not added by pragmatics on the basis of context. I will argue that certain other modals are performative in this sense.

 Nuyts (2001a) uses the term "performative" in a different, though related, way. For Nuyts, a performative use of a modal involves the speaker's commitment to the evaluation being made. Thus, for Nuyts (169) is performative because the speaker is committed to the addressee actually being under an obligation to leave. Any utterance which is performative in my sense will be performative in his as well, but the converse does not hold, since I might be committed to the claim that you are under an obligation while not doing anything more than asserting that such an obligation exists.

We can compare the two concepts of performativity by looking at example (170):

(170)   You must pray every day.

Nuyts can classify (170) as performative if the only speech act performed by the priest who utters it is to assert the existence of an obligation, so long as he is fully committed to the actual existence of this obligation. According to my sense of the term, however, this example counts as performative only if the semantics of *must* implies that some additional speech act takes place. At first glance, this example might seem to favor Nuyts's view. The priest would not say that his utterance of (170) imposes the obligation. As far as the priest is concerned, it's God who imposes the obligation, not him. However, this appealing argument is flawed because the non-assertive speech act which leads to performativity need not be the same in every case. All we are committed to is that the sentence with *must* performs an additional speech act. Note that not all imperatives are used to impose an obligation. For example, if the priest utters the following imperative, it has a function very similar to (170):

(171)   Pray every day!

We might describe the speech-act function of this imperative as "urging." Crucially, it does not make an assertion nor does it impose an obligation. Thus, we can describe *must* in this example as performative in my sense, as well as in Nuyts's.

4.1.2 Other classifications

There are many ways to classify sentential modals. The most common ones we find in the literature are a two-way distinction between epistemic and root modality and a three-way distinction among epistemic, deontic, and dynamic modality. The first of these is useful when one is trying to understand differences between epistemic modals and all other modals, but it is not appropriate when one is equally concerned—as we are here—with differences within the non-epistemic class. Therefore I will avoid using the term "root" in this sense, saying "non-epistemic" instead.

The second system is close to the classification that I will use, but differs from it in two ways. First, in terms of terminology, the label "deontic" is ambiguous between referring to the entire class of priority modals, as I have called it, and the subclass encompassing only those priority modals concerned with right-and-wrong, that is, ethical, moral, and legal norms. Since it will be necessary to talk about both of these types of modality in a clear way, I prefer to use the term "priority modal" for the broad class, and "deontic" for the narrower one. And, second, the traditional classification does not have a place for quantificational modals; since I believe that this group is important, it enters the classification.

I would also like to compare my system of classification to those of two other important theoretical works: Brennan (1993) and Hacquard (2006). Brennan's system is quite similar to mine, but there are some differences. First, she groups all non-epistemic modals into a single root class. I, in contrast, don't see evidence that all of the non-epistemic modals share anything except for not being epistemic. Thus, it makes no more sense to have a "root" class than it would to have, say, a separate class for all non-dynamic modals. Second, she does not group the volitional dynamic and quantificational modals together; instead, she has major classes of dynamic (equivalent to my volitional) and quantificational modals. And, third, she uses the term "deontic" in the broad traditional sense, and therefore is not clear about the various differences of meaning within this class. My introduction of the term "priority" solves this problem.

Hacquard has an innovative taxonomy for modals. She argues for a main three-way distinction among epistemic, "true deontic," and root modals. According to Hacquard, true deontics (which she also calls "ought-to-be" deontics, Feldman 1986) are those which place an obligation on the addressee; in other words, the true deontics are performative deontics in the sense introduced in Section 4.1.1. Epistemics and true deontics share the important property of being related to a participant in the speech situation, the speaker in the case of epistemics and the addressee in the case of true deontics. Other deontics (also called "ought-to-do" deontics, Feldman 1986) are grouped with the other root modals. Among the root modals, she recognizes two

TABLE 4.1. Semantic classifications for modality

				Root		
Traditional {	Epistemic |		Deontic		| Dynamic |	X
	Epistemic |					
My terms {	Epistemic |	Priority			Dynamic	
		Deontic	Bouletic	Teleological	Volitional	Quantificational
Brennan {	Epistemic |		Deontic	Root	Dynamic	Quantificational
Hacquard {	Epistemic |	True deontic |	Goal-oriented	Root |	Ability	X

types: goal-oriented modals (including ought-to-be deontics) and ability modals. She does not discuss quantificational modals at all.

Hacquard's proposal is important, but it suffers from an unwillingness to provide orthogonal classifications. That is, because she recognizes the importance of the performative subclass of deontics and its relation to epistemics, she separates them entirely from the non-performative deontics. This way of describing things is problematical, because the very same modal expression can be performative in some syntactic contexts and not performative in others. For example, Ninan (2005) argues that English deontic *must* is performative in root sentences, but not in embedded ones:

(172)   (a)   You must leave now.
        (b)   It is surprising that you must leave now.

As part of its semantics, example (172a) places a requirement on the addressee to leave. In contrast, while (172b) implies that you are required to leave, it does not itself impose this requirement. My system of classification labels both of uses of *must* as "deontic," thus allowing me to capture what they share, but recognizes that the former has a performative meaning which the latter lacks. We will discuss the performative semantics of deontic modals in Section 4.3 below.

These classifications of sentential modality are summarized in Table 4.1.

4.1.3 Modal force

The vast majority of work on sentential modality assumes that modals come in at least two strengths, strong and weak, or, as they are called in approaches inspired by modal logic, necessity and possibility. As we've seen, most theories of modality which are based on possible worlds treat necessity modals as universal quantifiers and possibility modals as existential quantifiers (or something similar to universal and existential quantifiers, as in the version of Kratzer's semantics which doesn't make the limit assumption). However, it is worth pointing out that there has been a recent challenge to the assumption that weak modals differ from strong ones in this way. Rullmann et al. (2007), building on work by Klinedinst (2007), suggest that weak modals in some languages (they discuss St'át'imcets) actually represent universal quantifiers with a contextually variable restriction on their domain of quantification.

According to Klinedinst, the semantics of a weak modal involves existential quantification over sets of accessible worlds, and then universal quantification over that subset:

(173) $[\![\,\Diamond\phi\,]\!]^{w,R} = \exists X[X \neq \emptyset \wedge X \subseteq \{w' : R(w, w')\}$
$\wedge\, \forall v[v \in X \rightarrow [\![\,\phi\,]\!]^{v,R}]]$

This meaning is equivalent to the standard analysis in modal logic, since the set X could contain just a single world. Rullmann et al. (2007), however, make an important change: rather than existentially quantifying over the set of accessible worlds, they claim that weak modals make reference to a specific set of accessible worlds.

Assume that a contextually given choice function F picks out of any set of worlds X a subset of X, and that R is an accessibility relation as usual. Rullmann et al.'s analysis can be expressed as follows:

(174) $[\![\,\Diamond\phi\,]\!]^{w,F,R} = \forall v[v \in F(\{w' : R(w, w')\}) \rightarrow [\![\,\phi\,]\!]^{v,F,R}]$

Their analysis is focused on a number of morphemes which can be described as evidentials, and so evaluating this theory in detail would require discussing the relationship between evidentiality and modality. We'll return to this topic briefly in Section 5.3. However, since they claim that evidentials, at least in this language, are epistemic modals, it is worthwhile to consider here their key idea, namely, that weak modals involve universal quantification with a contextually given "specific" subset of the set of accessible worlds as the domain of quantification.

Rullmann et al.'s analysis uses a choice function to restrict the set of accessible worlds to a contextually relevant subset. It does not make use of an ordering source (even though they call R a "modal base"). It would be possible for them to have the same semantic analysis simply by adopting Kratzer's theory and employing the ordering source to identify the relevant subset of the set of worlds compatible with the modal base.[4] That is, (174) is equivalent to Kratzer's semantics for a necessity modal under the limit assumption, since, for any particular function F, one can identify an ordering source which would lead to the same set of accessible worlds. Since both a choice function and an ordering source are contextually determined, there is no obvious pragmatic difference between the two approaches. Therefore it seems to me that Rullmann et al.'s analysis could be reformulated as follows: St'át'imcets evidentials are necessity modals, and the range of ordering

---

[4] This point was brought to my attention by a student in a semantics seminar, Joo Yoon Chung.

sources compatible with them is different from those familiar from analyses of English, German, and other languages.

4.1.4 Syntactic representation

While there is a vast literature on the syntax of modal expressions, we can distinguish several themes which are especially relevant to differences among the semantic categories of sentential modality. One such issue, to be discussed in more detail in Sections 4.3.2 and 4.4.1, is whether epistemic and non-epistemic modals differ in their argument structure. A recurring idea in generative syntax has been that epistemic modals take a single propositional arguments, like raising predicates, while non-epistemic modals take two arguments, an individual and a proposition, like control predicates. This analysis has been disputed by Bhatt (1998) and Wurmbrand (1999), but, as we will see in Section 4.3.2, they make the mistake of assuming that all non-epistemic modals must have the same analysis. The fact is that the raising/control distinction is an important one, but it does not line up with the epistemic/non-epistemic contrast in a neat way.

Another significant theme in the syntax literature is that different semantic categories of modals are located in different positions in syntactic structure. The most basic claim of this kind is that epistemic modals reside higher in the tree than non-epistemic ones (e.g., McDowell 1987; Brennan 1993; Marrano 1998; and Jackendoff 1972 has much relevant discussion). Hacquard (2006) modifies this claim, arguing that the relevant distinction is between epistemic and performative deontic modals on the one hand, and non-performative deontic and dynamic modals on the other. Based on extensive cross-linguistic evidence and an examination of both adverbs and functional morphemes, Cinque (1999) argues that there are many different positions for functional elements, including modals, each position being associated with a specific category of meaning.

There are two main types of evidence which have been produced for the proposal that different types of modals reside in different syntactic positions. First, there are a variety of arguments based on scope; we will review many of the relevant facts in discussing the non-truth-conditional analysis of epistemic modals in Sections 4.2.1–4.2.2. The idea here is that epistemic modals tend to take wide scope over other operators because they are higher in the tree to begin with. And, second, Cinque looks at cross-linguistic generalizations in the ordering of elements which express modality and other functional meanings, arguing that these generalizations can be explained in terms of different base positions for semantically distinct classes of elements. While many syntacticians accept Cinque's ideas, they have been criticized as well (e.g., Ernst 2001). There seems to be wide agreement in the syntax literature at this point that some modals stand higher in the structure than others, but it is not clear whether very many categories like those of Cinque need to be distinguished, or whether two or three are sufficient. The older scope-based arguments only support a smaller number, whereas Cinque's innovative cross-linguistic approach suggests more.

4.2 Issues in the semantics of epistemic modality

Epistemic modals have been the most well-studied of any group of modals. There are probably two reasons for this: first, an understanding of epistemic modals is helpful if one aims to explore the nature of knowledge, and this is an important concern within philosophy. And, second, a number of puzzles have been discovered which suggest that epistemic modals do not fit into standard theories of semantics. As a result, scholars have developed a variety of ways of thinking about their meaning, more or less radically diverging from mainstream semantic theory.

We find two main themes in arguments that epistemic modals do not fit into standard theories of semantics:

1. As pointed out in section 3.3.3, Lyons (1977) argues that subjective epistemic modals do not contribute to the ordinary, truth-conditional semantics of sentences containing them, but rather affect the type of speech act performed by the sentence. Many other authors take a similar position. I label this position, discussed in Section 4.2.1, the NON-TRUTH CONDITIONAL ANALYSIS. In the end, our conclusion will be that none of the arguments show that epistemic modals fail to contribute to truth conditions; rather, they show that epistemic modals do something in addition to contributing to truth conditions. Sections 4.2.2–4.2.3 explore the nature of this "extra" component of meaning.

2. Several philosophers have argued that the truth or falsity of an epistemic modal sentence is relative to the perspective of the person evaluating it. This implies that I may correctly describe

some utterance as true, while you may equally correctly say that it is false. This RELATIVISM about epistemic modality will be discussed in Section 4.2.4. Our conclusion in this section will be that the arguments for relativism fail. However, the arguments for relativism bring out an important point: we cannot define in a simple way whose knowledge is involved in the interpretation of an epistemic modal. The identity of this individual (or sometimes individuals) varies based on both grammar and pragmatics.

The overall conclusion one should draw from the discussion of epistemic modals in this chapter is that their semantics and pragmatics is much more complex than predicted by the semantic theories discussed in Chapters 2–3. Their meaning has a truth-conditional component which is well explained by sophisticated versions of possible worlds semantics, but there is more to it. Much of what falls outside of the truth-conditional semantics concerns discourse meaning.

4.2.1 The non-truth conditional analysis

A clear and concise recent discussion of the hypothesis that epistemic modals do not contribute to truth conditions is given by Papafragou (2006). Based on the literature, she identifies two phenomena which present puzzles for analyses of epistemic modality which say that it contributes to ordinary truth-conditional semantics.[5]

The first pattern concerns limitations on the embedding of epistemic modals. When a subjective epistemic modal is embedded in an *if* clause, a factive position, or under a verb of telling, the result is often odd or difficult to interpret (from Papafragou 2006, (8)–(10)):

(175) (a) ?If Max must/may be lonely, his wife will be worried.
       (b) ?It is surprising that Superman must be jealous of Lois.
       (c) ?Spiderman told me that Superman must be jealous of Lois.

In contrast, as pointed out by Lyons (1977), when the modal is interpreted objectively the embedding is much easier (Papafragou's (33), (38)):

[5] This literature includes Leech (1971); Jackendoff (1972); Lyons (1977); Palmer (1986, 1990, 2001); McDowell (1987); Nuyts (1993, 2001b); Westmoreland (1995); Garrett (2000); Drubig (2001); Faller (2002, 2006b); and von Fintel (2003); later additions include Yalcin (2007a,b); Stephenson (2007); Portner (2007a); and von Fintel and Gillies (2007b).

(176) (a) If Paul may get drunk, I am not coming to the party.
       (b) It is surprising that the victim must have known the killer.
       (c) The police told reporters that the victim must have known
            the killer.

(In light of the considerations given by Nuyts 2001b, discussed above
in Section 3.3, these cases might be better described as intersubjective
rather than objective; Papafragou uses the term "objective," however.)

     This contrast is not limited to modal auxiliaries. As pointed out by
Lyons, expressions like *possibly* most naturally receive a subjective interpretation, while *it is possible that* is more easily interpreted as objective:

(177) (a) ?If Paul will possibly get drunk, I am not coming to the
            party.
       (b) If it is possible that Paul will get drunk, I am not coming
            to the party.

This contrast reinforces the conclusion that the difference between (175)
and (176) is due to the subjective/objective distinction.

     The contrast between (175) and (176) can be explained if one assumes
that subjective epistemic modals do not contribute to the ordinary,
truth-conditional semantics of the sentence. These are contexts in
which the compositional semantics calls for something with truth conditions, i.e. a proposition. If a subjective modal is present, something
goes wrong. On the non-truth-conditional view, what goes wrong is
that the sentence containing the modal does not have truth conditions.

     A weakness in this argument for the non-truth-conditional analysis
is that subjective epistemic modals can be embedded in other contexts,
for example:

(178) (a) Mary believes that Max must be lonely.
       (b) There can't have been a mistake. (negation scopes over
            modal, from von Fintel and Gillies 2007b, (12))

Under standard analyses, the arguments of *think* and negation should
be a proposition, and so it's unclear why the subjective modal can be
embedded under them.

     A related argument for the claim that epistemic modals do not con-
tribute to truth conditions is given by Westmoreland (1998: (3.21)):

(179) (a) We know a large amount of "dark matter" must be there.
       (b) We know a large amount of "dark matter" is there.

Examples (179)a–b are virtually equivalent, and thus Westmoreland says that "The syntactic argument of *know* is MUST Φ, but the semantic argument is the content of Φ" (p. 80). This type of example is interestingly different from those we saw above. In (175)–(177), the claim was that epistemic modals cannot be embedded because they lack truth conditions; Westmoreland's claim is that they can be embedded, but don't contribute to the semantic argument of the predicate which embeds them. The difference between these arguments should be troubling for anyone who believes that epistemic modals do not contribute to truth conditions—in general, does the lack of truth conditions lead to ungrammaticality or invisibility in subordinate clauses? Another relevant point is that (179) clearly represents the objective use of an epistemic modal, and so it would be difficult to fit this pair of examples into Lyons's way of thinking about modality.

The example in (179), and some others given by Westmoreland, are difficult to evaluate for another reason. Because the sentences begin with *we know...*, they are about the knowledge of the speaker and addressee. But *must* itself is about the knowledge of the speaker or of the speaker and addressee. That is, we have one epistemic operator embedded under another very similar one. For this reason, the facts noted by Westmoreland can be explained by principles of modal logic.[6] Recall that, in basic modal logic (namely T), we have that $\Box\Box\phi$ entails $\Box\phi$. (Remember that T is $\Box p \to p$; just substitute $\Box\phi$ for p to get $\Box\Box\phi \to \Box\phi$.) In S4, we have the additional axiom 4 ($\Box p \to \Box\Box p$), and so $\Box p \leftrightarrow \Box\Box p$ in S4. Thus, the Simple Modal Logic Hypothesis leads us to expect that (179a–b) are equivalent, assuming that the epistemic accessibility relation is by its nature reflexive and transitive.[7]

Let us try to reformulate Westmoreland's argument based on better examples. The problem of equivalence between $\Box\Box p$ and $\Box p$ might not arise if the two $\Box$'s are different, for example, if one is based on the speaker's knowledge and the other on someone else's knowledge.

---

[6] A similar point was made by Geurts and Huitink (2006), though they focus on cases where the two operators are clause-mates and so do not address Westmoreland's argument directly.

[7] As pointed out by Steve Kuhn, this argument also leads us to expect that both sentences are equivalent to *A large amount of dark matter must be there*. However, this sentence seems to imply that our knowledge is indirect, a fact which has been used as an argument that *must* is an evidential and not a modal. See Section 4.2.2 for discussion. In this respect, it fits into the general perspective endorsed by Westmoreland. Nevertheless, the implication of indirectness is not relevant to Westmoreland's argument based on (179).

Therefore, one strategy for constructing better examples would be to use a third-person subject instead of *we*:

(180)   Mary knows that a large amount of "dark matter" must be there.

However, this strategy does not work. In this example, the meaning of *must*, like that of *know*, is based on Mary's knowledge, not the speaker's. That is, we still have the $\Box\Box p$ structure. More generally, when *must* is part of the argument of a sentence-embedding verb, it is sensitive to the matrix subject's knowledge or beliefs. We'll return to this point in Section 4.2.4.

Westmoreland's argument was not improved by changing the matrix subject, and so, as a next step, we might try changing the matrix verb. In some cases, the result is strange:

(181)   ?Mary hopes/doubts that a large amount of "dark matter" must be there.

With other matrix verbs, as in (178a), the resulting sentences are acceptable, and so we must figure out what they mean. Consider first (182a):

(182)   (a)   Mary thinks it may rain.
        (b)   It is compatible with Mary's beliefs that it will rain.
        (c)   Mary thinks it will rain (but she's not sure).

Example (182a) has a meaning very similar to (182b). If the two are equivalent, this fact can probably be explained in terms of possible worlds semantics. For example, $\Box\Diamond p \leftrightarrow \Diamond p$ is a theorem of several modal logics which can be build on the standard axioms, including the one with axioms **KD4E**, a reasonable approximation of the logic of belief. Assuming that *may* expresses a modality based on Mary's beliefs, we predict that the two examples are equivalent.[8] However, there is a feeling that (182a) is actually closer in meaning to (182c), that is, that it makes a somewhat stronger statement than (182b). If it is indeed stronger, there is no simple explanation for this fact in terms of modal logic.

We see a related problem in (183):

---

[8] Yalcin (2007a) also observes this pattern and argues that it supports his semantics for epistemic modality (based on the analysis in dynamic logic, see Section 3.2.3 above). Recall that he analyzes $\Diamond$ as a test and that he hypothesizes that attitude verbs shift the information state with respect to which their complement is interpreted. Thus, in (182a), we'd find that *may* tests whether the rain is compatible with Mary's beliefs.

(183)   (a)   I pray that God may bless you. (Palmer 1990)
        (b)   I pray that God will bless you.

As pointed out by Portner (1997), (183a–b) are nearly equivalent; but, in contrast to the previous case, this example involves deontic (or some other kind of priority) modality, and again modal logic provides no simple explanation. Moreover, it is likely that the *may* in (183a) is not simply a deontic modal. It rather appears to be the optative *may* seen in sentences like *May God bless you!* In order to understand this form of *may*, we must consider discourse modality.

The term HARMONIC MODALITY is used in cases in which two modal elements are of similar meaning. (For relevant discussion, see Halliday 1970; Lyons 1977; Coates 1983; Palmer 1986, 1990; Portner 1997, among others.) In some cases, each of the two modals seems not contribute a separate modal meaning, but rather they reinforce one another, a situation which has been called MODAL CONCORD (see Geurts and Huitink 2006; Zeijlstra 2008). All of the examples in (179)–(183) can be described as harmonic in a general sense, and those in (182)–(183) might exhibit modal concord. We will introduce some of the issues relevant to harmonic modality and modal concord in Section 5.3.

Thus far in this section, we have discussed arguments based on embedding for the hypothesis that epistemic modals lack truth conditions. Papafragou discusses another argument for the same idea. This one concerns patterns of assent, dissent, and questioning towards sentences containing epistemic modals. Consider the following dialogue containing a subjectively interpreted epistemic modal:

(184)   A: Max must be lonely.
        B: That's not true./ I agree./ Are you sure?

We understand B's response as targeting the proposition that Max is lonely. For example, with *I agree*, B agrees that Max is lonely, not that it follows from A's knowledge that Max is lonely. In contrast, when the modal is interpreted objectively, it is possible to target either the modal proposition or the prejacent with which it combines:

(185)   A: The victim must have known the killer.
        B: That's not true./ I agree./ Are you sure?

The response *I agree* can mean that B agrees that the evidence supports the conclusion that the victim knew the killer, or that B agrees that the victim in fact knew the killer.

If we assume that subjective epistemic modals do not contribute to truth conditions, it is easy to explain this contrast. When B says *I agree*, he is agreeing with some salient proposition. The proposition that Max is lonely is salient because it is the meaning of the phrase under the scope of the modal; moreover, because *must* does not contribute to truth conditions, there is no other salient proposition. That is, there is no proposition "must+p" which B can agree with.

I have some qualms about the data in (184). There is a difference between *Are you sure?* and *I agree*. I think that B's response *Are you sure?* does mean "Are you sure that Max is lonely?", as predicted. However, the predictions do not seem quite right for *I agree* (and *That's not true* as well). Obviously, *I agree* does not mean "I agree that *you* are of the opinion that Max is lonely," yet it doesn't quite mean "I agree that Max is lonely" either, as pointed out by Faller (2002). Rather, B is saying that his assessment of how likely it is that Max is lonely matches A's. In other words, both A and B are making modal claims, but A is talking about A's subjective assessment while B is talking about B's. This way of seeing the example makes it rather similar to the facts which have been taken to support relativism; we'll look at those arguments in detail in Section 4.2.4.

Overall, then, it seems that both a meaning incorporating the epistemic modal (subjective or objective) and one not incorporating it can be targeted in conversation. This point weakens the argument that subjective modals do not contribute to truth conditions. But note that we have not proven that they do contribute to truth conditions. One might still agree with Lyons that they do not contribute to truth conditions, operating at the speech-act level instead; in connection with this idea, one could claim that *I agree* targets a speech act, rather than a proposition. That is, B is agreeing that *Max must be lonely* is a reasonable thing to say. If a defender of Lyons were to take this approach, she would have to explain the pragmatics of *agree* in much more detail, since (as pointed out by Steve Kuhn, p.c.), there are many kinds of speech acts to which one cannot say "I agree," for example, acts of asking, christening, and joining two people as husband and wife.

We have seen that the arguments for the claim that epistemic modals lack an ordinary truth-conditional semantics are weak. Moreover, there is evidence that they do contribute to truth conditions. In order to have clear predictions about what proposition is expressed by a sentence or what speech act is performed, we need a precise theory of discourse meaning. Within possible worlds semantics, the standard such theory is that of Stalnaker (1974, 1978). Let's take a look at the behavior of epistemic modals in light of this theory.

As discussed in Section 3.2.1, Stalnaker introduces the notion of the common ground and gives an account of assertion based on it. Recall that the common ground is the set of propositions mutually presupposed by participants in the conversation, and that to assert a proposition is to offer it for inclusion within the common ground. The assertion of a proposition is successful if the proposition does enter the common ground as a result of the assertion.

In light of this view of assertion, consider the following examples from Papafragou (2006: (20)):[9]

(186)   (a)   My grandfather must be sick.
        (b)   My grandfather may be sick.
        (c)   My grandfather is sick.

Uttering any of these results in an assertion. It is most clear in (186c) what has been asserted: that the speaker's grandfather is sick. But uttering (186a) would certainly result in some information being added to the common ground: that the speaker has good evidence that her grandfather is sick, or something similar. In terms of the model of discourse, this means that a modal proposition has been asserted. A similar point goes for (186b).

If we adopt Stalnaker's model of discourse, there's no way to avoid the conclusion that examples like (186a–b) involve assertion. If one wishes to avoid this conclusion, one must give an alternative model of discourse meaning. Let us briefly consider two analyses which do just that: dynamic logic and the probabilistic semantics of Swanson (2006a, 2006b, 2007).

The issue of truth conditions in dynamic logic

As we saw in Section 3.2, there is a clear sense in which dynamic logic does not assign truth conditions to the epistemic modal operators. In particular, $\Diamond\phi$ acts as a test on an information state, so that $s[\Diamond\phi]$ is either s or $\emptyset$, depending on whether s is compatible with ϕ. This update potential cannot be reproduced in terms of a static truth-conditional semantics for $\Diamond\phi$ and a simple rule of assertion. In the next few pages,

---

[9] Papafragou makes this argument and several related ones, but her presentation is not completely explicit because it does not reveal the necessary assumptions about the nature of discourse semantics.

we will look at how the arguments for the non-truth-conditional analysis look from the perspective of dynamic logic.

Another fact about dynamic logic makes it especially relevant for thinking about the issue of truth conditions. Within this theory, an information state s is typically understood to be a belief state or knowledge state, not a discourse-oriented entity like the common ground or context set. (It could be otherwise. There's nothing in the logic itself which prevents it from being used to provide a model of conversational update.) This focus on mental states rather than shared information suggests a connection with the concept of subjective modality. Here we will focus on what dynamic logic can teach us about the nature of subjective epistemic modality.

As we saw in Section 3.2, one of the most important features of dynamic logic is that it can help us understand the status of examples like (187):

(187)   (a)   It might be raining outside. ... It isn't raining outside.
              (Groenendijk, Stokhof, and Veltman 1996, (1))
        (b) $\Diamond p \wedge \neg p$

This example is quite close to the dissent patterns discussed above. A more natural dialogue version of (187), with changes meant to emphasize a subjective reading, is the following:

(188)   A: The baby might be cold. ($\Diamond p$)
        B: No, he isn't. ($\neg p$)

Dynamic logic is just a logic, but Groenendijk et al. (1997) make some proposals concerning the rules of conversation which would govern the use of sentences with the meanings assigned by dynamic logic. The basic idea is that you should only say what is supported by your own belief state, with the intention that your hearers update their belief states for what you say. Hearers should either comply by updating their belief states or, if doing so would result in the absurd state, object.

In terms of these (admittedly simplistic) principles, we can think about (188). In order for A to say $\Diamond p$, his beliefs must be compatible with the baby's being cold. B should either update her own belief state for $\Diamond p$ or object that she is unwilling to do so. Because the update potential assigned to $\Diamond p$ is just a test, the only reason B would object is if her belief state is not compatible with p, and by saying $\neg p$, this is just what B does. A should then update his belief state for $\neg p$ or object that he is unwilling to do so. Let's assume that A's beliefs are compatible with the baby's not being cold; he will then accept what B said, and both A and B will end up with the belief that the baby isn't cold. Note that A would not be able to sincerely say $\Diamond p$ again (as we saw in Section 3.2, $\neg p \wedge \Diamond p$ is inconsistent).

Example (184) was used to argue that subjective epistemic modals do not contribute to truth conditions on the grounds that it is difficult to target the meaning of the sentence including the modal for assent or dissent. Dynamic logic can explain this fact in an interesting way. There is nothing wrong with a formula like $\neg \Diamond p$. Let's examine the following:

(189)   A: The baby might be cold. ($\Diamond p$)
         B: No, he can't be. ($\neg \Diamond p$)

The update potential for $\neg \Diamond p$ is given as follows:

 (i) $s[\neg \Diamond p] =$
 (ii) $s - s[\Diamond p] =$
 (iii) (a) $(s - s) = \emptyset$, if s is compatible with $V(p)$;
   (b) $(s - \emptyset) = s$, if s is not compatible with $V(p)$.

B says *He can't be* ($\neg \Diamond p$) because her belief state is not compatible with the baby's being cold. In other words, she believes that the baby is not cold. So far so good. But A should then update his belief state for $\neg \Diamond p$. We know that A's belief state is compatible with p because he said $\Diamond p$. Therefore, updating for $\neg \Diamond p$ results in the absurd state (see line (iii)(a)). Therefore, $\neg \Diamond p$ is not a cooperative way for B to make her point that she disagrees with what A said. Intuitively, what goes wrong is that B is telling A "update your belief state with *He can't be cold*," even though it's clear that, as far as A's beliefs go, he could be cold.

Put together, (188)–(189) show that, within dynamic logic, it makes sense to respond to a modal statement $\Diamond p$ by denying p, but it doesn't make sense to respond by denying $\Diamond p$. This is very similar to the pattern observed with subjective epistemic modals in (184). (Keep in mind that we are discussing only the subjective interpretation of $\neg \Diamond p$. On an objective interpretation, this response is clearly acceptable and, in that case, perhaps receives an entirely different semantics, as proposed by Lyons.)

We should consider too whether dynamic logic can also say something interesting about fact that it is difficult to embed in a subjective epistemic modal. Let's consider an example with a conditional, (175a), repeated here:

(190)   ?If Max must/may be lonely, his wife will be worried.

Here is how we calculate the update potential for $\Diamond p \to q$:

(i) $s[\Diamond p \to q] =$
(ii) $s[\neg(\Diamond p \land \neg q)] =$
(iii) $s - s[(\Diamond p \land \neg q)] =$
(iv) $s - s[\Diamond p][\neg q] =$
(v) (a) $s - s[\neg q]$, if s is compatible with $V(p)$;
   (b) $s - \varnothing[\neg q]$, if s is not compatible with $V(p)$
(vi) (a) $s - (s - s[q])$, if s is compatible with $V(p)$;
   (b) s, if s is not compatible with $V(p)$
(vii) (a) $s \cap V(q)$, if s is compatible with $V(p)$;
   (b) s, if s is not compatible with $V(p)$

According to this semantics, $\Diamond p \to q$ performs a $\Box$-like test for $\neg p$. That is, if s already contains the information $\neg p$, we get s back. However, in contrast to a simple test of $\Box\neg p$, the result of failing this test is not $\varnothing$, as would be the case with $\Box\neg p$, but rather a gain of information q. (Since $\Diamond p \to q = \neg(\Diamond p \land \neg q) = (\neg\Diamond p \lor q) = (\Box\neg p \lor q)$, this meaning is to be expected.) There's nothing wrong with this meaning, and so dynamic logic as currently formulated is unable to explain the embedding facts. This failure is not surprising, since it works with an unrealistic analysis of conditionals based on standard logic and does not say anything about other embedding structures.

Dynamic logic provides an important perspective on the non-truth-conditional theory. It shows that it is possible to accept the view that epistemic modals do not contribute to truth conditions without following Lyons's proposal that they have a special meaning involving speech acts. In dynamic logic, update potentials are basic, while truth conditions are defined in terms of them. Assertion is a special case of update: an update is an assertion if it's made using a sentence which has truth conditions. A modal sentence has an update potential just like any sentence; its lack of truth conditions makes it special, but not fundamentally different from other sentences. However, dynamic logic as it stands does not explain all of the important facts which motivate the non-truth-conditional perspective. In particular, it does not account for the embedding phenomena. In order to have a hope of doing so, it would need to be combined with more sophisticated analyses of embedding structures such as those presented in other theories that take a dynamic perspective on meaning (e.g., Heim 1982, 1982; Kamp and Reyle 1993; Frank 1996; and Yalcin 2007a, 2007b). Of course, there's no guarantee at this point that it would be successful at explaining all of the relevant facts even with such improvements.

The issue of truth conditions in a probabilistic semantics

Swanson (2006a,b, 2007) proposes a semantic theory according to which the meaning of sentences is based on a subjective probability.[10] The details are complex, so I will only sketch them here (in particular leaving out the complexities having to do with NP quantification). Let us start with the type of meaning a sentence has: according to Swanson, a sentence denotes a set of functions from propositions to probabilities (numbers between 0 and 1).[11] For example, a non-modal declarative sentence has a relatively simple meaning:

(191) $[\![$ *It is raining* $]\!] = \{ \{ \langle \{w : \text{it is raining in } w\} \rightarrow 1 \rangle \} \}$

We can think of this meaning as identifying a single proposition, that it's raining, and associating it with the maximum probability 1.

When we add an epistemic modal, in effect we switch the 1 to a wider range of probabilities. For example, with *might*, we link our proposition to all of the probabilities which reach some threshold μ. Suppose μ is .4 and we don't list probabilities beyond one decimal place:

(192) $[\![$ *It might be raining* $]\!] =$
$\{ \quad \{\langle \{w : \text{it is raining in } w\} \quad \rightarrow .4 \rangle\},$
$\{\langle \{w : \text{it is raining in } w\} \quad \rightarrow .5 \rangle\},$

$\ldots$

$\{\langle \{w : \text{it is raining in } w\} \quad \rightarrow 1 \rangle\} \quad \}$

We can think of this meaning as specifying that it is OK to assign "it is raining" any probability greater than or equal to .4.

It is possible to define a notion of "true" which lets us describe a non-modal sentence as true or false. A sentence is "truth-apt" (Swanson 2007: 33) iff you only find 1's when you look inside the set it denotes. Thus, (191) is truth-apt. A truth-apt sentence is true in w iff all of the propositions which are assigned probability 1 by functions in its

---

[10] As pointed out in Section 3.1.3, Yalcin (2007a,b) also sketches a semantics for probability operators as part of a general analysis of epistemic modality. His approach is much more standard in that it assigns truth conditions to all sentences with epistemic modals and probability operators, and so it doesn't have anything special to say about the question of whether sentences with epistemic modals have truth conditions.

[11] As in standard possible worlds semantics, propositions are sets of possible worlds. Technically, he represents all sets with characteristic functions, but I'll simplify the explanations by using sets.

denotation are in fact true in w. For example, (191) is true in the actual world iff it is raining in the actual world. In contrast, (192) is not truth-apt. Thus, it cannot be assigned a truth value. But even though (192) is neither true nor false, it has a perfectly clear meaning, the set of probability assignments given. Unlike dynamic logic (which also has a perfectly explicit notion of non-truth conditional meaning), Swanson's theory is not based on the notion of update potential. That is, this theory is static, not dynamic.

Swanson's theory also has something to say about how sentences which do not have truth conditions can be embedded in such contexts as under *believe*.

(193) $[\![$ *Mary believes that it might be raining* $]\!]$ =
            $\{\,\{\,\langle\{w : \text{Mary is disposed to conform to (192) in } w\} \rightarrow 1\rangle\,\}\,\}$

What does it mean to be "disposed to conform to" a set of probability assignments? Swanson isn't very explicit about this, but we can get an idea by thinking about our example. Suppose that Mary is pretty insistent about taking an umbrella, but is unwilling to cancel plans to play golf. The pattern of actions makes sense if we assume that she assigns the proposition that it is raining a probability which is not too high and not too low. Maybe somewhere between .4 and .6. In that case, she is disposed to conform to (192).

Despite the fact that it elegantly accounts for the intuition that epistemic modals lack truth conditions, this theory faces some challenges. One is the fact that it's not clear how this theory can distinguish embedding structures that allow subjective epistemic modals, like *believe*, from those that do not. For example, subjective *might* does not occur easily under *know*, and it's hard to see why this should be. Moreover, it's not obvious how to express the difference between objective and subjective epistemic modality; the theory makes epistemic modality fundamentally subjective, so perhaps he'd say that the so-called objective cases really don't represent epistemic modality at all.

Another problem is the fact that Swanson's theory cannot be fitted into an existing model of discourse semantics. Swanson is quite aware of this issue, in particular the fact that it cannot be integrated into Stalnaker's model of assertion. Recall that to assert a proposition, according to Stalnaker, is to add it to the common ground. In the probabilistic semantics, however, not all sentences can be said to express a proposition. In particular, a non-truth-apt one like (192) cannot be.

This means that we must reassess the role of the common ground in discourse semantics.

One move that Swanson could make is to abandon the notion of common ground altogether. Although he doesn't do this, he does give it a diminished role. He proposes that the fundamental function of a declarative sentence is to offer advice about what the addressee's belief state should be like. For example, if I say *It is raining*, your belief state should assign probability 1 to the proposition that it's raining. In other words, I advise you to make your belief state conform to (191). Likewise, if I say *It might be raining*, I advise you to conform it to (192). In other words, I advise you to assign the proposition that it's raining some probability greater than or equal to μ (.4 in our example). This advice is not the same thing as assertion, according to Swanson, because it is not an effect on the common ground.

An issue that obviously arises at this point is what it means to advise someone to do something. Advising is a kind of speech act and has something to do with the semantics of imperatives, which are easily used to give advice (e.g., *Take an umbrella!*). To make clear predictions about meaning in discourse, Swanson would have to offer an explicit theory of advising.

Besides offering advice on what to believe, an utterance also has some effect on the common ground. When the sentence does not contain a modal, this effect should be as in Stalnaker's original theory: in the case of (191), the proposition that it is raining should join the common ground. (Generally, if the meaning of a sentence is a singleton set containing just one function, and that function assigns some proposition to 1, that proposition should become part of the common ground.) But if the sentence has an epistemic modal or other marker of uncertainty, it's not so clear what should happen. Swanson offers a few ideas: for example, in the case of (192), the common ground should not entail that it won't rain. But he does not provide much more than this.

Almost incidentally, Swanson mentions that there is a third component to the meaning of *might p*. Sometimes when someone says *might p*, this results in the hearer's coming to see the possibility that *p*, a possibility that may have been overlooked before. For example, the baby keeps crying. We've given him milk, changed his diaper, and tried to put him to sleep. Nothing works, and we have no idea what the problem could be. Then I have a realization and say:

(194)   Oh, he might be cold.

Now that we see the possibility, we know what's wrong!

Swanson provides a model for this change of coming to be aware of a possibility.[12] This model involves assigning an individual two probability spaces: a fine-grained one which assigns probabilities to every proposition which is relevant to his actions, and a coarse-grained one which only assigns probabilities to a subset of these, namely, those he is aware of. When someone comes to be aware of a possibility, the coarse-grained space becomes a little less coarsely grained. Moreover, Swanson claims that the the two-probability-space model is needed to explain a component of the "force" (Swanson 2007: 44) of *might*. Though he doesn't say so explicitly, this reference to force implies that raising a possibility is a kind of speech act.

At this point, it is abundantly clear that Swanson's theory takes epistemic modals to be performative in the sense defined above. Sentences containing them perform three speech acts:

1. They have some kind of effect on the common ground; that is, they assert something.
2. They advise the addressee on how to update her subjective probabilities.
3. They raise a possibility of which the addressee might have previously been unaware.

I am not sure whether we need all three of these as independent speech acts. Perhaps the second and third acts can be combined into a single piece of advice: "conform your coarse probability space to $[\![\phi]\!]$." (Note that you can only conform your coarse probability space to $[\![$ *He might be cold* $]\!]$ if you are aware of the possibility that he is cold.) But in any case, Swanson's work is very interesting in that it identifies these three types of speech act which might be relevant to epistemic modals. We'll return to the issue of whether epistemic modals are performative in Sections 4.2.3.

Subjectivity

We have referred to and used the concept of subjectivity many times, but have not yet thought explicitly about what the correct definition of this notion is. Lyons's view is that subjective modals do not contribute to truth conditions and that root sentences containing them are not

---

[12] Yalcin (2007a,b) also has a theory of the difference between a proposition which is merely compatible with your beliefs and a proposition which you are aware of and which is compatible with your beliefs.

used to make assertions. Instead, the modal causes the sentence to perform a different kind of speech act: it "qualifies the illocutionary act" (Lyons 1977: 805). But Lyons does not make explicit what the nature of the illocutionary act is, and he does not integrate any of his ideas into a general theory of discourse meaning.

As discussed in Section 3.3.3, Kratzer (1981) also makes a proposal concerning the nature of subjectivity. She proposes that a subjective interpretation arises in German when the ordering source contains superstitions or other "non-objective" assumptions. However, while it's clear that superstitions cannot count as objective in German culture, her discussion does no more than provide a single example of subjectivity. We need a more general statement of what makes an ordering source subjective. Moreover, even if we had such a statement, it's not clear how subjectivity in this sense could explain the assent/dissent and embedding facts.

Tancredi (2007) argues that what has been called subjective epistemic modality is actually doxastic, that is based on the speaker's beliefs, not knowledge. His argument concerns (195), his (1c):

(195)   That may be Jones.

Alice knows that Jones has been invited to the party and believes (wrongly) that he is out of the country and so cannot make it. Someone knocks on the door. If Alice says (195) on an epistemic reading, according to standard accounts (e.g., modal logic or Kratzer's theory) what she says should be true, since that he has been invited is knowledge, but that he's out of the country is just a belief. Yet, as Tancredi points out, it "would be an inappropriate thing for Alice to utter in a context where only her own state of mind is at issue" (p. 3).[13] Tancredi takes this as an argument that the semantics of (195) does not concern just what Alice knows, but rather everything she believes. However, his argument assumes that if the sentence is true, it is automatically an appropriate thing for Alice to say, but truth is not enough to guarantee appropriateness. The Quality Maxim of Grice (1975) makes this point clearly. According to the Maxim, a speaker should only say something if she believes it to be true and if she has adequate evidence for it. Assuming that Alice takes her belief that Jones is out of the country to be knowledge, she neither believes (195) to be true nor has adequate

---

[13] Tancredi states that the sentence is false, but I don't share this judgment. The feeling that it is false might be due to the fact that it is similar to an eavesdropper case of the kind discussed in Section 4.2.4.

evidence. Hence, it is not an appropriate thing for her to say, even if it is true.[14]

Papafragou (2006) analyzes subjectivity within Kratzer's theory in terms of subtypes of epistemic conversational backgrounds. She proposes that an epistemic modal involves quantification over the members of a relevant group of "knowers." In each context, there is a group of people whose knowledge is relevant; an epistemic possibility sentence $\Diamond p$ is true iff for every member of the group, there is some world compatible with what they know where p is true. The following definitions, based on the system of Section 2.3.5, reformulate the proposal given by Papafragou (p. 1694):

(196)   **Accessibility relation function** (Papafragou style)
        A is an accessibility relation function iff:

        1. Its domain is a set of actual and/or hypothetical contexts of utterance; and
        2. Its range is a set of sets of of accessibility relations.

(197)   **Epistemic accessibility relation function** (Papafragou style)
        A is an epistemic accessibility relation function iff:

        1. A is an accessibility relation function; and
        2. For every context c in the domain of A and every accessibility relation R, $R \in A(c)$ iff, for some x in the group of people relevant to the interpretation of the modal in c, R is the relation which holds between two worlds w and w' iff everything which x knows in w is also true in w'.

(198) $[\![\, \Diamond \beta \,]\!]^{w,c,\langle\langle W,A\rangle,V\rangle} = 1$ iff for for every $R \in A(c)$, there is a v such that $R(w,v)$, and $[\![\, \beta \,]\!]^{v,c,\langle\langle W,A\rangle,V\rangle} = 1$

In this system, sentences are interpreted with respect to sets of accessibility relations, one for each member of the set of relevant knowers.

---

[14] Tancredi also says "If Alice has no basis for making a distinction between those of her beliefs that constitute knowledge and those that do not, then she has no way of ever knowingly making a subjective epistemic modal statement at all based on her own epistemic state." I am not sure what he means by a "basis for making a distinction." One can certainly make a judgment about which of one's beliefs constitute knowledge; in the extreme case, one may take all of one's beliefs to be knowledge, or one can take some to rest on a foundation which does not justify calling them knowledge. I don't see why the way one makes these judgments should affect the issue at hand. The theory of epistemic modality certainly doesn't rely on speakers having a reliable way to distinguish what they know from what they wrongly believe. Veltman's comment quoted in Chapter 3, footnote 23, is relevant here.

For example, if John and Mary are the relevant people, then *It may rain* is true iff it is compatible with John's knowledge that it rains, and it is compatible with Mary's knowledge that it rains. A similar definition is discussed and rejected by von Fintel and Gillies (2007a, (12)).[15]

Within this framework, Papafragou defines subjectivity as a case in which the speaker is the sole member of the group of relevant individuals. In contrast, an objective interpretation occurs when the broader community is relevant; the meaning is roughly *Given what each person knows, ...* In between are various intermediate cases, for example when the speaker, hearer, and a few others in the room are relevant. She also says that "the main difference between subjective and objective epistemic modality is that the former, but not the latter, is *indexical*, in the sense that the possible worlds in the conversational background are restricted to what the *current* speaker knows *as of the time of utterance*" (p. 1695).

There are problems with Papafragou's way of working out her ideas.[16] First, her definition implies that the knowledge of every member of the community must be counted separately. Papafragou asks us to consider an objective interpretation of (199), spoken by a meteorologist, where the relevant evidence is "observable evidence available to his professional community" (p. 1695):

(199)   It may rain tomorrow. (Papafragou 2006, (27))

But the definition in (198) does not depend on the evidence available to the community as a whole; rather, it depends on the evidence in the possession of each member of the community individually. Thus, (198) does not exactly match the meaning Papafragou has in mind. Moreover, the definition runs into problems. If one particular meteorologist fails to have access to a crucial piece of data (his internet connection is down), according to Papafragou's definition, (199) would be false. This

---

[15] Von Fintel and Gillies reject their version of (198) because the modal does not quantify over a single set of accessible worlds: "But now we have lost the idea that a modal is a quantifier over a modal base: there is no one set of possibilities throughout which we check for some ϕ-worlds" (p. 8). We can make this abstract objection a bit more concrete by noting that there is no evidence that *must* or *may* need to refer to multiple accessibility relations when they have non-epistemic readings.

[16] The first problem mentioned in the text only arises in the form stated if the modal is also dependent on an ordering source. Otherwise, the lack of information by an individual could never make a possibility statement false. However, assuming that epistemic modals depend on both a modal base and a doxastic or stereotypical ordering source, the problem does arise. Moreover, one can re-create the problem with a necessity modal or with negation plus a possibility modal, independently of whether we have an ordering source.

seems incorrect. Papafragou would have to say that this meteorologist is excluded from the community. A definition in terms of the evidence available to the community as a whole would work better than the one in (198).

The second problem concerns the relationship between the two ways Papafragou suggests we look at subjectivity. On the one hand, she says that a modal is subjective when {speaker of c} constitutes the relevant group. On the other hand, she says that subjective modals are indexical while objective ones are not. However, a modal can be objective in the sense that the relevant group includes more people than the speaker while still being indexical in that it depends on what the speaker knows. This is obvious in the case when the relevant group is {speaker of c, addressee of c}, and even in the meteorologist case Papafragou implies that the speaker is one of the individuals whose knowledge is relevant. This implies that even objective interpretations are indexical, in Papafragou's sense. Strictly speaking, the use of a modal should count as objective for Papafragou only if the speaker's knowledge is not relevant; it is unclear whether such cases exist, though see von Fintel and Gillies (2007a) for a suggestion that they do.

A third problem for Papafragou is that her definition implies that $\Diamond\beta$ will be true if β is compatible with each individual's knowledge, even if everyone's knowledge is not jointly compatible with β. To see this, note the scope relations in (198): the definition says that for every individual x, there is a world v compatible with x's knowledge where β is true. It is not required that there is any single v compatible with every individual's knowledge where β is true. Suppose that the phone rings, and we wonder whether it is one of our children calling. I know it's not Noah, but think it could be Ben. You know it's not Ben, but think it could be Noah. See Figure 4.1. Consider (200):

(200)   That might be one of our children.

According to Papafragou's definition, (200) is true on the group-knowledge reading, but this prediction seems incorrect. The problem arises because (198) takes the sets of propositions which are possible for all of the individuals in the group, and intersects these sets; a proposition is possible for the group if it is in this intersection. Since one of my children might be calling is compatible with my knowledge, and is compatible with yours, it might be true for us as a group. Note that von Fintel and Gillies (2007a) offer a better definition, their (16), which gets this case right by taking the sets of possible worlds accessible for

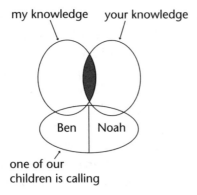

FIGURE 4.1. The group knowledge situation for (200)

each individual, intersecting these sets, and saying that a proposition is possible for the group if some world in this intersection makes the proposition true. According to this definition, a proposition is possible for the group in Figure 4.1 iff it intersects the gray area. The proposition that one of our children is calling is not compatible with what we jointly know in this sense.

Papafragou argues that her analysis can explain the embedding and assent/dissent facts outlined above. With regard to embedding in conditionals, she says:

> If subjective epistemic modal expressions are indexical (in the sense of the previous section), this restriction is explained. The environment inside the antecedent of a conditional cannot be an environment in which the speaker performs a mental evaluation of a proposition with respect to her belief-set. (Papafragou 2006: 1696)

(The relevant data is in (175)–(176).) However, I do not see how this explanation works. Papafragou's theory of subjective modality does not develop a concept of "performing a mental evaluation of a proposition." It only offers truth conditions based on an indexical accessibility relation. She suggests that her theory does more than this by drawing a parallel with performative verbs, for example (201), saying "crucially, both subjective epistemics and performatives are tied to the here-and-now of the conversational exchange" (p. 1696):

(201)   This is Emma's sister, I conclude. (Papafragou 2006, (31b))

However, being tied to the here-and-now of the conversational exchange does not make for performativity. Nor does it provide an analysis of "performing a mental evaluation." It is nothing more than indexicality; and there's no problem with putting an indexical into an *if* clause.

Papafragou explains the assent/dissent facts in terms of two ideas. (The relevant data is in (184).) One of these is more relevant to the issues which will be discussed in Section 4.2.4, and we will return to it there. The other is the point that the accessibility relation for a subjective epistemic modal is *externally inscrutable*. That is, the statement concerns the only speaker's private beliefs, and nobody else can be more expert on what the speaker's beliefs are than the speaker himself. Therefore, one cannot challenge the truth of a subjective epistemic statement. But one can challenge the proposition expressed by the sentence minus the modal. This is what we see in (184). This reasoning correctly predicts that one can challenge one's own subjective statement, as in (202):

(202)   Clark Kent may be Superman. No, that's not right. Clark Kent *must* be Superman. (Papafragou 2006, (42a))

I believe that Papafragou's analysis here is essentially correct.

We should also recall Nuyts's (2001b) ideas about subjectivity, outlined above in Section 3.3. Like the other authors discussed above, he believes that subjectivity has something to do with the nature of the evidence which supports an epistemic statement. Moreover, he clearly distinguishes the two issues which must be kept separate: (i) the quality of evidence and (ii) the status of the evidence as shared vs. held by the speaker alone. Either (or both) of these could be relevant to subjectivity. Papafragou's ideas focus on (ii). Nuyts does not provide a formal theory, but he also favors (ii), based on data from Dutch and German. He thus proposes that subjectivity is properly opposed to intersubjectivity, not objectivity.

Nuyts makes another intriguing point. He argues that subjectivity, whether characterized in terms of (i), (ii), or a combination of the two, should be thought of as an evidential component of meaning. Like the classical evidentials which indicate direct evidence, inference, or hearsay, it allows the speaker to indicate something about the nature of the evidence on which he is basing his contribution to the conversation. A number of other authors have suggested a link between evidentiality and epistemic modality, and we turn to their ideas in the next subsection.

In my opinion, Papafragou's and Nuyts's discussions of subjectivity suffer from their failure to keep in mind Kratzer's finding that modals are sensitive to two (possibly more) information sources, the modal base and ordering source. We do not have to choose between a theory based on the (inter)subjectivity of evidence and one based on the quality of evidence. The modal base concerns knowledge; this could be the knowledge of the speaker alone or a group including the speaker. The ordering source brings in other considerations besides what is firmly known, including the speaker's beliefs and stereotypical expectations; these considerations can be judged according to how reliable they are. Therefore I suggest two notions of subjectivity, one based on the modal base and one on the ordering source:[17]

(203) (a) **Subjectivity$_{MB}$ vs. Intersubjectivity**

   (i) A modal base f is epistemic with respect to context c iff there is a group G whose knowledge is relevant in c and which includes the speaker of c such that, for any world w, $f(w) =$ the set of propositions known jointly by G in w.

   (ii) Suppose f is an epistemic modal base with respect to context c. Then f is intersubjective iff G contains more than one individual. Otherwise, f is subjective$_{MB}$.

   (iii) The use of an epistemic modal in context c is intersubjective iff it is interpreted with respect to an intersubjective modal base. Otherwise, it is subjective$_{MB}$.

(b) **Subjectivity$_{OS}$ vs. Objectivity**

   (i) A modal base g_1 is more subjective (less objective) than another g_2, relative to a context c, iff members of the community C relevant in c tend to hold the kind of information represented by g_1 in lower regard than that represented by g_2.

   (ii) A use u_1 of an epistemic modal in context c_1 is more subjective than another use u_2 of the same modal in context c_2 iff the two uses are made relative to the same modal base, and u_1 is made relative to a more subjective ordering source than u_2.

Subjectivity$_{MB}$ is a categorical concept: A statement is objective if it is based on knowledge shared within a relevant group; otherwise, it is subjective. In contrast, subjective$_{OS}$ is scalar: One statement is judged to be more subjective or objective than another. Though one could define a concept of "maximal objectivity" (or "minimal subjectivity"),

---

[17] Subjectivity$_{OS}$ is based on Kratzer's discussion; this formulation is very rough, at best.

I don't know of any need for such a concept in the analysis of natural language.

A final interesting discussion relevant to subjectivity is given by von Fintel and Gillies (2007a). They suggest that the group of individuals whose knowledge contributes to the meaning of an epistemic modal can contain a non-human knowledge source. They consider the example (204) (their (9), due to Hacking 1967):

(204)   The hulk might be in these waters.

The scenario is that the mate of a salvage ship is working from accurate logs and charts to determine where to dive for sunken treasure. He confidently asserts (204), but he has made a mistake in his calculations. The correct location is some miles to the south. In this case, many speakers view (204) as false, and von Fintel and Gillies argue that this is the case because we consider the logs and charts to count as members of the group whose "knowledge" is relevant to the the modal's meaning; more precisely, the logs and charts represents accurate information, and this accurate information can play the role of a person's knowledge in the semantics of an epistemic modal. Since the logs contain information which rules out that the hulk is in these waters, (204) is false. Though they do not say so explicitly, it appears that they consider a situation of this kind to represent objective epistemic modality. This approach has some similarity to (203b), but it is formulated in terms of the modal base rather than the ordering source. Though they do not focus on these issues in a clear way, we can build on von Fintel and Gillies's ideas to create a picture according to which objectivity, subjectivity, and intersubjectivity are three independent dimensions for classifying modal bases: A modal base is objective if G contains a reliable non-human knowledge source; it is subjective if G contains the speaker; and it is intersubjective if G contains humans other than the speaker.

I hope that this discussion helps to give some theoretical content to the concepts of subjectivity, objectivity, and intersubjectivity. A variety of different ideas have been developed, and this is an area of active research, so I expect there to be further progress in understanding these issues in the near future. It remains unclear what exactly we should make of the assent/dissent facts and the embedding facts with which we started the discussion. Papafragou gives a plausible analysis of the the fact that it is difficult to assent to or dissent from a subjective epistemic modal, and her idea can be carried out in terms of the notion of subjectivity$_{MB}$ in (203) or von Fintel and Gillies's approach. We still have not seen an adequate theory of the embedding facts given in (175)–(176). We'll return to Lyons's idea that epistemic modals are performative, and that performativity has something to do with the mysteries of subjectivity, in Section 4.2.3.

4.2.2 Evidentiality

Some linguists have argued that evidentials and epistemic modals are the same thing. This idea can be phrased in one of two ways: One can say that epistemic modals are evidentials, or that evidentials are epistemic modals. These two ways of describing the idea are not equivalent. Someone who claims that epistemic modals are evidentials typically assumes that we more or less understand the semantics of modals, but that epistemics are not semantically modals, and belong with the class of evidentials instead. Authors who take this perspective include Westmoreland (1995, 1998) and Drubig (2001). In contrast, a linguist who proposes that evidentials are epistemic modals typically thinks that we understand the semantics of modals pretty well, and that the right way to think about evidentiality is to classify it as a subtype of modality. For example, Matthewson et al. (2006) and McCready and Ogata (2007) take this perspective. In this section, we'll briefly discuss the first perspective: if we assume that modality and evidentiality are distinct, are words like *must*, traditionally called modals, actually a kind of evidential? We will return to evidentiality again briefly in Section 5.3, but a serious treatment of the second perspective will have to await another occasion (in particular, Portner, forthcoming).

The claim that epistemic modals are evidentials has a weak and a strong version. The weak version says that they incorporate evidentiality as part of their meaning, but may be different from pure evidentials. The strong version says that they are just evidentials, not modals at all. The weak version is supported by Nuyts (2001a) and von Fintel and Gillies (2007b), while the strong version is argued for by Westmoreland (1998) and Drubig (2001). It is actually a bit difficult to say whether the evidence supports the weak or strong version, because those who hold the strong version make specific assumptions about what the right semantic analyses of evidentials and modals are: they believe that modals quantify over possible worlds and contribute to truth conditions, while evidentials affect the type of speech act which the sentence performs, and do not contribute to truth conditions. Thus, in discussions of modals and evidentiality, it is sometimes assumed that if some element contributes to truth conditions, it is a modal, and if it doesn't, it is an evidential. However, we have seen that some scholars believe that epistemic modals (at least the subjective ones) do not contribute to truth conditions, while others believe that evidentials, like epistemic modals, do (Matthewson et al. 2006; McCready and Ogata 2007). Related to this confusion is the fact it is not settled whether we should use the term "evidential" in a descriptive or theoretical sense. If it a descriptive term, it could be that different evidentials require different theoretical analyses (some as speech-act operators, others as modals, for example). If it's a theoretical term, one will try to arrive at a correct semantic analysis of evidentiality, and then use this to decide whether a particular morpheme is a modal, an evidential, or something else. Because of the unsettled nature of the questions being asked, we must be very cautious as we evaluate the arguments presented for and against the view that epistemic modals are evidentials.

Arguments that epistemic modals are evidentials often focus on English *must, might,* and *may*. There are three types of argument that these words are evidentials: First, we have the array of facts which suggest more generally that some epistemic modals do not contribute to truth conditions (see Section 4.2.1). This claim is relevant if we assume that evidentials do not contribute to truth conditions but modals do. Second, we have arguments based on scope properties. And, third, we have the fact that the semantics of epistemic modals often involves concepts which are similar to those associated with evidentials, such as inference, indirect evidence, or hearsay.

We have already discussed the first argument, the claim that epistemic modals lack truth conditions, and have not found convincing evidence for it. The second argument involves scope properties: it is widely known that epistemic modals tend to take wide scope over other sentential operators, including negation, other modals, and tense (examples from Drubig 2001; see also Groenendijk and Stokhof 1975; Enç 1996; Abusch 1997):[18]

(205)   (a)   John may not be at home.
         (b)   John may have to be available for consultation.

---

[18] All of the arguments which aim to show that epistemic modals do not contribute to truth conditions involve scope as well, in that they involve the claim epistemic modals cannot occur in the semantic scope of certain operators. However, they differ from the cases discussed in this section in being more directly relevant to the issue of truth conditions.

(c) Anti-ci   re'an-aha-kon.   (Garo, Sino-Tibetan)
     market-to   go-PAST-PROBABLE
     'He probably went to the market.'

Example (205a) aims to show that epistemic modals always take scope over negation; (205b) exemplifies the fact that epistemic modals invariably scope over deontic ones, and (205c) illustrates a similar pattern with respect to tense.

Drubig's claim about negation has been challenged by Papafragou (2006) and von Fintel and Gillies (2007b). Some relevant examples are the following:

(206)   (a)   John cannot be at home. (Drubig 2001, (7)b)
       (b)   It's not possible that John is at home.

Drubig claims that (206a) involves a different kind of epistemic modality from that we are familiar with, but does not provide a clear argument for this point. Thus I think we should conclude that epistemic modals can take scope under negation.

The scope facts involving other modals and tense are more promising. I don't know of any examples in which a non-epistemic modal has scope over an epistemic one. As for tense, von Fintel and Gillies (2007b) claim that tense can have scope over epistemic modals:

(207)   The keys might have been in the drawer. (von Fintel and Gillies 2007b, (15))

This sentence has a meaning close to "Based on the evidence that I had in the past, it was possible that the keys were in the drawer." (It also has a meaning with the expected scope, "Based on the evidence I have now, it is possible that the keys were in the drawer.") The authors think that the first meaning is derived by means of a past tense operator with scope over the modal. However, matters are not so simple. Putting past tense over the modal would result in one of the following two representations:

(208)   (a)   PAST(might(keys in drawer))
       (b)   PAST(might(PERF(keys in drawer)))

In (208a), I assume that the perfect verb form *have been* represents the past tense. Thus, the material in the scope of *might* is tenseless. Given that *might* allows the forward-shifting of temporal reference (so that (209a) concerns my present knowledge about a future possibility), we should be able to say something like (209b).

(209)   (a)   The keys might be in the drawer (tomorrow)
        (b)   The keys might have been in the drawer today.
        (c)   He might have been available yesterday/next month.
              (Condoravdi (2002), (34a))

Condoravdi (2002) claims that on a metaphysical (historical) interpretation, *might have* allows this kind of forward-shifting, as in (209c), but not on an epistemic interpretation. So the question is whether (209b) is acceptable with a past tense epistemic meaning.[19] To my mind, it is not, but the possibility of a metaphysical interpretation makes it difficult to say for sure. We'll return to the issue of the temporal interpretation of modals in Section 5.1.1. Until we understand the complexities of the semantic relationships among modality, tense, and aspect, it is too soon to say whether an epistemic modal can have scope under tense.

Based on the discussion of (205)–(207) we can conclude that epistemic modals do not freely take narrow scope. (The precise details are still a bit unclear.) However, this fact does not show that they are evidentials. The scope behavior could simply be a result of syntax. As we saw in Section 4.1.4, some syntactic analyses hold that epistemic modals occupy a position quite high in the clause, and by itself this predicts that they will tend to have wide scope. Moreover, the fact that epistemic modals do not always have wide scope is difficult to account for if one believes both that they are evidentials and that evidentials have to take widest scope.

The final argument for the claim that epistemic modals are evidentials comes from the fact that the meanings of epistemic modals often seem to be based on concepts like direct evidence, indirect evidence, reported information, and hearsay. These concepts are familiar from studies of evidentiality (see Willett 1988). Along these lines, we've already discussed Nuyts's (2001b) claim that (inter)subjectivity is a kind of evidential category. Many authors have pointed out that English *must* seems to require that the evidence relevant to the modal statement be indirect (e.g., Karttunen 1972; Kratzer 1981; Veltman 1986; Stone 1994; Westmoreland 1998; Drubig 2001; von Fintel and Gillies 2007b). Thus, it's odd to say (210) if one is standing under the rain:

(210)   It must be raining.

---

[19] It is acceptable on a present tense interpretation where the perfect form indicates "unreality" or "counterfactuality." See Section 5.1.1 for discussion. And Condoravdi might say that it is acceptable on a metaphysical interpretation.

Example (210) is felicitous when the speaker has inferred that it is raining, not when he has directly observed it.

Also relevant are reportative modals, for example, German *sollen* and Dutch *moeten* (relevant recent work includes Mortelmans 2000; de Haan 2001; and Faller 2006b):

(211)   Eine Zinssenkung   soll   unmittelbar
        A   interest-rate-reduction MODAL directly
        bevorstehen.
        approaching
        'A reduction in interest rates is said to be imminent.' (Ehrich
        2001)

It is very common for evidential systems to contain forms which express that some statement is made based on indirect evidence or on someone else's report. Thus *must* and *sollen* show that epistemic modals may base their meanings on the same intuitive concepts as evidentials. This fact, in turn, could be taken to show that some epistemic modals are evidentials, at least in the weak sense.

The fact that *sollen* encodes reported information does not prove that it is not a modal. In terms of Kratzer's theory, we could describe its meaning by saying that *sollen* requires a conversational background consisting of facts which the speaker knows as a result of what someone told him. (This "reportative" conversational background would probably function as an ordering source.) Likewise, we could describe the "inferential" component of *must*'s in terms of Kratzer's semantics. There are at least two approaches one could follow. First, one could say that the modal base for *must* requires a set of facts known through means other than direct observation. Or, second, one could require that the ordering source come into play, as in (212):

(212)   (i) $[\![$ must S $]\!]^{c, f, g}$ expresses that S is necessary with respect
               to c, f, and g.
        (ii)   Presupposition: $[\![$ must S $]\!]^{c, f, \emptyset}$ is not true in the actual
               world.

Part (i) agrees with Kratzer on the semantics of *must*; part (ii) says that the epistemic modal base by itself does not establish that S is true. Thus, the truth of *must S* depends on the information in the ordering source. Since the ordering source for *must* is typically doxastic or stereotypical, this means that the speaker bases his statement partially on his beliefs or rules of thumb, not just on known facts.

At this point, it is important to recall some of the cautionary notes raised at the beginning of this section. One of these concerned the uncertain status of the word "evidential." Is it a descriptive or a theoretical term? It is reasonable to conclude, based on (210)–(211), that epistemic modals can function as evidentials in a descriptive sense. But this leaves open the question of whether their meanings should be analyzed in terms of precisely the same theory as uncontroversial examples of evidentials. Given that we have not discussed the proper analysis of evidentials, we are not in a position to answer this question. Another point to keep in mind is the difference between the strong and the weak version of the hypothesis that epistemic modals are evidentials. We have seen that the arguments for the strong version fail. These arguments were based on the assumptions that evidentials take wide scope over other sentential operators and do not contribute to truth conditions. Therefore, we are left with four possibilities:

1. Epistemic modals are in fact evidentials in the strong sense, but we don't have good evidence for this yet.
2. Epistemic modals are evidentials in the strong sense, but this is so because all evidentials are in fact modals.
3. Epistemic modals are evidentials in the weak sense; they incorporate components of the meaning of evidentials, but they also have modal components of meaning lacked by pure evidentials.
4. Epistemic modals and evidentials are descriptively similar, but are not related to one another at a theoretical level.

At this point, we can ignore possibility 1, as it currently has no evidence to back it up. The other three possibilities remain reasonable options to consider. Section 5.3 contains a brief discussion of current approaches to evidentiality.

4.2.3 Performativity

I defined a modal as performative if its presence in a declarative sentence results in the sentence performing, as part of its conventional meaning, a speech act different from, or in addition to, the usual speech act of assertion. Lyons (1977) endorses performativity for subjective epistemic modals when he proposes that they modify the speech act associated with the sentence, so that a sentence containing such a modal performs a weaker speech act than assertion. Unfortunately, neither Lyons nor other linguists with similar ideas (e.g., Halliday 1970; Palmer 1986; Bybee and Fleschman 1995b) make clear the precise nature of this speech act. Moreover, as we have seen in the discussion of subjectivity, it seems that even sentences with subjective epistemic modals assert something. Therefore, if these modals are performative, they cause the sentence to perform two (or more) speech acts, assertion and something else besides.

We have already seen that the theory of Swanson (2006a, 2006b, 2007) is explicitly performative. He proposes that, in addition to an assertion, a sentence with *might* does two other things: it advises the addressee on how to modify her beliefs, and it raises a possibility of which the addressee might not have been previously aware. As far as I know, Swanson's theory is the only one to propose two non-assertive speech acts in this way, but I think we can see each of them individually as being closely related to other scholars' ideas about the performativity of epistemic modals. Let us discuss each in turn.

First, we have the advice speech act. It seems plausible to me that when previous authors such as Lyons say that a sentence containing an epistemic modal performs a speech act weaker than assertion, they have in mind something similar to Swanson's advice. Of course, as mentioned earlier, Swanson is not explicit about what it is to offer advice, and so it's not possible to draw this connection in detail. Nevertheless, the idea behind Lyons's proposal is that, when someone utters a sentence with an epistemic modal, she is trying to affect the addressee's beliefs, just as with ordinary assertion, but is doing so in a way which is consistent with her own uncertainty about what the facts are. For example, consider (213):

(213) (a) The dog is hiding.
       (b) The dog might be hiding.

According to classical speech-act theory (e.g., Austin 1962; Searle 1969), when you assert (213a), you are trying to get the addressee to believe that the dog is hiding.[20] In terms of probabilities, to fully believe p is to assign it probability 1, and so to assert (213a) is to try to get your addressee to assign it probability 1.

It is reasonable and responsible to try to get your addressee to assign p probability 1 if you yourself assign it this probability, but suppose you don't. In that case, the reasonable thing to do is to try to get your addressee to assign it a probability similar to yours. For example, if

[20] In particular, you are trying to get the speaker to believe that the dog is hiding through the particular Gricean mechanism at the heart of speech act theory. See especially Searle's work for details.

I assign it the probability .5, I could use (213b) to try to get you to assign it a probability above .1. (I could then say some other things, like *but it's not all that likely*, to make sure you don't assign it too high a probability.) What do we call this act of trying to get the addressee to assign a probability above .1? Is it an assertion? Perhaps we don't want to call it that, since assertion might be understood as implying certainty. (To tell the truth, I think it would be equally reasonable to call it an assertion.[21]) If we don't call the speech act which involves trying to get someone to assign a probability above .1 an assertion, Swanson's *might* is indeed a modifier of the speech act, and we can see a link between his ideas and the speech act-modifier view of Lyons.

The other kind of speech act discussed by Swanson is that of raising a possibility. An utterance of (213b) can raise the possibility that the dog is hiding. Swanson represents raising a possibility in terms of the addressee's belief state: If the addressee was not aware of the possibility that the dog is hiding, his coarse-grained belief state did not assign a probability to it. But after the speaker raises the possibility, his coarse-grained belief state will come to assign it a probability. (If the addressee takes the speaker's doxastic advice, it will have a probability greater than or equal to μ, .1 in our example.)

Other scholars have had similar ideas about the semantics of epistemic modals. For example, Yalcin (2007a,b) introduces the concept of a "modal resolution" for an agent, a partition of the set of possible worlds such that the agent does not recognize distinctions which cut across the cells of the partition. Yalcin's idea is somewhat similar to Swanson's, but rather than employing two probability functions, it employs a single one in combination with a mechanism (the resolution) for keeping track of which possibilities the agent is aware of.

A somewhat similar idea is proposed by von Fintel and Gillies (2007b) when they suggest that, in saying (213b), one is both asserting a modal proposition and "proffering (with an explicit lack of conviction)" that the dog is hiding or "giving advice not to overlook the possibility" that the dog is hiding (von Fintel and Gillies 2007b: 13). Their proposal is novel in that it suggests that "proffering" may be the

---

[21] C. Davis et al. (2007) outline a model of discourse semantics which allows us to say that a declarative sentence always asserts something, but that the assertion is always made with respect to a contextually determined minimum probability, the QUALITY THRESHOLD. Their idea is aimed at evidentials, and it appears that they believe evidentials and epistemic modals to be at least somewhat different. Nevertheless, their idea does seem potentially useful for the problem under discussion here.

right way to look at the performativity of epistemic modals; however, they are not explicit about what it is to proffer or give advice. Similarly Portner (2007a) proposes that when one uses a sentence containing an epistemic modal, one both asserts a modal proposition (same as von Fintel and Gillies here) and makes the non-modal proposition a shared possibility. For example, in the case of (213b), the speaker asserts that there is a possibility that the dog is hiding and makes the proposition that the dog is hiding a shared proposition.

The concept of sharing a possibility seems to be similar to von Fintel and Gillies's "proffering," but Portner (2007a) attempts to explain it in a more precise way. In particular, I propose a model of sharing a proposition which is formally similar to asserting it. Within Stalnaker's model, assertion can be modeled either in terms of the context set or the common ground. Portner uses the latter type of formulation: the update potential of a declarative sentence is to add the proposition it expresses to the common ground. We can formalize this idea as follows:

(214)   **Assertion:** For any declarative sentence ϕ, the update potential of ϕ used in context c, $[[\,\phi\,]]^c$, is defined as follows:

$$cg\,[[\,\phi\,]]^c = cg \cup \{[[\,\phi\,]]^c\}$$

Suppose that, in addition to the set of mutually presupposed propositions (the common ground), we have a broader set of shared propositions, the COMMON PROPOSITIONAL SPACE (CPS). The CPS is the set of propositions in which the participants in the conversation are mutually interested; for example, any proposition which is a candidate for inclusion in the common ground because someone has assertively uttered a sentence with that meaning, is in the CPS.

The state of a conversation is represented by (among other things) a pair consisting of the common ground and CPS: $\langle cg, cps \rangle$, where $cg \subseteq cps$. In these terms we can describe the update potential of a declarative sentence with *might* as follows:

(215)   **Update potential of *might*:** for any sentence ϕ of the form *might* ψ, the update potential of ϕ used in context c with modal base f and ordering source g, $[[\,\phi\,]]^{c,f,g}$, is defined as follows:

$\langle cg, cps \rangle [[\,\phi\,]]^{c,f,g} = \langle cg', cps' \rangle$, where

(i) $cg' = cg \cup \{[[\,\phi\,]]^{c,f,g}\}$, and

(ii) $cps' = cps \cup \{[[\,\psi\,]]^{c,f,g}\} \cup \{[[\,\phi\,]]^{c,f,g}\}$

This definition incorporates assertion: the modalized proposition $[[\,\phi\,]]^{c,f,g}$ is added to the common ground. But, in addition, two propositions are added to the common propositional space: the non-modal proposition $[\![\psi]\!]^{c,f,g}$ and the modalized proposition $[\![\phi]\!]^{c,f,g}$. (The latter must be in the CPS because it's been asserted.)

We can think of adding $[\![\psi]\!]^{c,f,g}$ to the CPS as one way of making precise the idea that a sentence with *might* "proffers" the proposition under the scope of the modal. This proposal has two important virtues: it is simple and builds upon Stalnaker's highly fruitful ideas about assertion. Nevertheless, we certainly need a better understanding of the CPS. See Portner (2007a) for further discussion.

In this section we have encountered a number of different ideas concerning the performativity of epistemic modals. These ideas can be classified into two main types:

(i) Analyses which claim that a sentence with an epistemic modal expresses a reduced level of commitment. Here we have Lyons's speech-act perspective and Swanson's probability-based account. It is interesting that there are parallels between both of these and theories of evidentiality: in the case of speech acts, Faller (2002, 2006c), and in the case of probablities, McCready and Ogata (2007) and C. Davis et al. (2007).

(ii) Analyses which say that sentences with epistemic modals bring up or proffer a possibility. Here we have quite a few different ideas. One useful way of classifying them separates the analyses of von Fintel and Gillies (2007b) and Portner (2007a) on the one hand from those of Swanson (2007) and Yalcin (2007a) on the other. The "proffering" approach of von Fintel and Gillies and Portner conceives of the performativity of epistemic modals as affecting the discourse context. In contrast, Swanson and Yalcin focus on an agent's belief state, and in particular on keeping track of the difference between propositions of which the agent is aware or unaware. These two approaches are parallel to the two ways of thinking about update potential in dynamic approaches to meaning: recall from Section 3.2 that dynamic logic typically conceives of an update potential as information-change potential, that is a function which affects someone's knowledge, while other dynamic approaches think of it as context-change potential, a function which updates the context set or common ground. Apart from this distinction between a belief/knowledge orientation and a conversational/contextual orientation, all versions of this approach to performativity agree

that the traditional notion of information state must be supple-
mented by a method for distinguishing propositions which are
fully accepted from those which are less completely integrated.

Overall, even if one finds attractive the hypothesis that epistemic
modals are performative, it is clear that the theories in this area must be
developed in greater detail. Future research is likely to focus on whether
epistemic modals are really performative, and, if they are, what the
nature of their performativity is.

4.2.4 Relativism

All semantic theories of modality acknowledge that the meaning of a
sentence containing a modal—or any other sentence, for that matter—
depends on the context in which it is used. For example, we have
seen that, within the possible worlds approach, the accessibility relation must be partially determined by context. Such theories have been
called CONTEXTUALIST (Egan et al. 2005). In recent years, a number of
philosophers have argued that sentences containing epistemic modals
depend for their literal meaning[22] not just on the context in which
they are used, but also on the context in which they are evaluated. The
hypothesis that the semantic meaning of a sentence depends on the
context of evaluation is called RELATIVISM. For example, suppose that
we accept a truth-conditional, possible worlds semantics for epistemic
modality. Within this mainstream perspective on modal semantics, the
relativist says that the set of possible worlds over which the modal
quantifies depends on the evaluator's knowledge. (It might depend on
the speaker's knowledge too, or on the evaluator's knowledge alone.)

Two main arguments for relativism have been offered, and once we
consider these, it will become more clear how a relativist semantics
should look. (There are a number of other arguments, but these are
either variants of the two main ones, or else related to points we have
already discussed above in Sections 4.2.1–4.2.2. For an extensive critique
of relativist theories of epistemic modals, see von Fintel and Gillies
2008.) One type of argument is closely connected to the assent/dissent
tests discussed above. The following examples are from MacFarlane
(2006: 5):

---

[22] It's not controversial that the pragmatic meaning of a sentence can depend on the
context of evaluation, provided that we apply the term "pragmatic meaning" in a relatively
liberal way. The relativist is making a claim about the semantic meaning of sentences, not
just the pragmatic meaning.

First case: You overhear George and Sally talking in the coffee line. Sally says, "I don't know anything that would rule out Joe's being in Boston right now" (or perhaps, more colloquially, "For all I know, Joe's in Boston"). You think to yourself: I know that Joe isn't in Boston, because I just saw him an hour ago here in Berkeley. *Question*: Did Sally speak falsely?

Second case: Scene as before. Sally says, "Joe might be in Boston right now." You think to yourself: Joe can't be in Boston; I just saw him an hour ago here in Berkeley. *Question*: Did Sally speak falsely?

MacFarlane thinks that the answer to the first question is "no" and that the answer to the second question is "yes." Let us call "you" in these cases the person who overhears a conversation not directed at him, the "eavesdropper." Many arguments for relativism are based on the presence of an eavesdropper.

If it's true that Sally spoke falsely in the second case, this suggests that the meaning of (216) depends on the context of evaluation.

(216)   Joe might be in Boston right now.

After all, it was compatible with Sally's knowledge that Joe was in Boston, and so (216) should be true if the meaning of the sentence depends only on the speaker's knowledge. The only justification for saying that she spoke falsely was the fact that the eavesdropper knows that he's not in Boston. In this way, the example provides support for a relativist semantics.

The other main argument for relativism concerns embedding structures. As Egan et al. (2005) develop an argument similar to MacFarlane's, they employ sentences in which an epistemic modal is embedded under a verb like *say*. I will avoid giving their whole argument here, and instead focus on the key issue of embedding structures like (217):

(217)   Sally says that Joe might be in Boston.

What's crucial about this sentence is that it seems to fit into a general pattern like the following:

(218)   (i)   A says S. [= (217)]
        (ii)   What A says is true.
        (iii) Therefore, S. [= (216)]

If we can apply this pattern to sentences with epistemic modals, and if what Sally said is true, we should be able to conclude that Joe might be in Boston. As Egan et al. (2005) show, this conclusion leads to an argument in favor of relativism.

I will consider this second argument first, since it is much weaker. The argument rests on the assumption that S has the same meaning in (i) as in (iii). However, when S contains an epistemic modal, it does not have the same meaning in these two contexts. For example, in (216) the meaning of *might* depends on the speaker's knowledge, while in (217), it depends on Sally's knowledge, not the speaker's. That is, (216) means something like "it is compatible with the speaker's knowledge that Joe is in Boston." (The semantics of modals is a bit more complex, of course, but this description is enough to make the point.) In contrast, (217) means something like "Sally says that it is compatible with her (Sally's) knowledge that Joe is in Boston."

We need a theory which predicts that an epistemic modal in an embedded sentence can depend on the knowledge of some individual other than the speaker. Stephenson (2007) makes an explicit proposal as to how this switch from speaker to another individual should be done. She proposes that the meanings of several kinds of sentences depends on the identity of a "judge." Such sentences include those containing epistemic modals and predicates of personal taste such as *delicious* or *fun*. We can incorporate this idea into a simple possible worlds semantics by adding the judge as a parameter of interpretation j, so that our interpretation function looks as follows: $[\![S]\!]^{w, \langle\langle W, R\rangle, V\rangle, j}$ [23]

(219) (a) **Epistemic accessibility relation**: R is an epistemic accessibility relation iff, for any individual j (the "judge") and world w in the domain of R, $v \in R(j, w)$ iff v is compatible with what j knows in w.

(b) **Meaning of *might***: $[\![\text{might } S]\!]^{w, \langle\langle W, R\rangle, V\rangle, j} = 1$ iff for some $v \in R(j, w), [\![S]\!]^{v, \langle\langle W, R\rangle, V\rangle, j} = 1$.

In the simplest cases, the judge parameter is identified with the speaker, but it can also be identified with some other individual. In particular, in our example it is identified with Sally.

Egan et al. (2005) consider an approach similar to Stephenson's, but they reject it on the grounds that epistemic modals behave very differently from other words for which it is plausible that there is some kind of extra semantic parameter whose value varies from case to case. They consider in particular words like *nearby* and *local*; they think *local*

---

[23] Stephenson brings the judge into the semantics by means of a special syntactic argument, PRO$_j$, which always refers to the judge. However, she mentions that it would also be possible to make the modal directly dependent on the judge parameter, as I do in the text.

means something like "in the neighborhood of *x*," where *x* can be specified in one of several ways by context:

(220) (a) John goes to a local school.
       (b) Every one of my classmates has a job at a local bar.

As pointed out by Mitchell (1986), in (220a) *x* can be John or the speaker. Example (220b) can mean that every classmate has a job at a bar which is close to where that classmate lives; in this case, *x* behaves as a bound variable. It is clear that the judge parameter proposed by Stephenson does not behave like *x*. In particular, in simple examples the judge cannot be identified with a third party, someone other than the speaker. For example, (216) cannot mean that it is compatible with Joe's knowledge that he is in Boston; nor can it mean that it is compatible with the Queen of England's knowledge. This counter-argument is not convincing, however, because it assumes that we have considered all of the various kinds of variable-like elements which exist in natural language. That is, they assume that, if the judge parameter is a variable-like element, it must either be a classic indexical (similar to *I*) or a simple variable (like *x*). But we know that there are many more kinds of variable-like elements in natural language than these. I will return to this point below.

Once we have incorporated the judge into our semantics, we need a theory which can predict who the judge is in particular contexts. Clearly, the judge can be the speaker, and examples like (217) show that it can also be identified with another individual mentioned in the sentence. But, as we have seen, the identity of the judge is not entirely free. Therefore, we must find some principles which tell us, in particular cases, who can be the judge. As far as I know, no scholar has yet undertaken a detailed study of this problem, but I think that we can identify two general kinds of principle which might be useful.

On the one hand, there may be grammatical principles operating at the syntax/semantics interface. Stephenson, for example, proposes that in the structure *A thinks/says that S*, the meaning of the verb causes A to be the judge for *S*. The semantic rule might look as follows:

(221) **Meaning of *says*:** $[\![\alpha \text{ says } S]\!]^{w,\langle\langle W,R\rangle,V\rangle,s} =$
       $[\![\text{says}]\!]^{w,\langle\langle W,R\rangle,V\rangle,s} ([\![\alpha]\!]^{w,\langle\langle W,R\rangle,V\rangle,s}, [\![S]\!]^{w,\langle\langle W,R\rangle,V\rangle,\alpha})$

She also suggests that *because* has a similar grammatical effect on the identify of the judge. On the other hand, there may be pragmatic principles; it seems to me that one can view relativism as the claim that only one pragmatic principle matters: the judge is the evaluator.[24] In my view, a combination of the two approaches is most likely correct: we need some grammatical principles of the kind offered by Stephenson and some pragmatic ones as well.

One idea worth pursuing is that the judge parameter behaves as a shiftable indexical of the kind discussed by Schlenker (2003) and Anand and Nevins (2004). For example, Anand and Nevins (2004: 31) discuss pronouns in Slave. The Slave first-person morphology refers to the speaker in unembedded contexts, but its reference may switch when embedded. In the following example, the first-person object of 'hit' is coreferential with the matrix subject of 'say' (Anand and Nevins's (38b)):[25]

(222)   Simon [rásereyineht'u] hadi.
        Simon [2.sg-hit-1.sg]   3.sg-say
        'Simon said that you hit him (=Simon).'

When the pronoun is embedded under an element glossed "want," it may be be coferential with the matrix subject, or it may refer to the speaker (their example (38a)):

(223)   John [beya   ráwoz'ie]   yudeli.
        John [1.sg-son 3.sg-will-hunt] 3.sg-want-4.sg
        'John wants his son to go hunting.' or
        'John wants my son to go hunting.'

In light of these few facts concerning shiftable indexicals, the behavior of English epistemic modals may not seem so strange. In both cases, when the shiftable element is in a complement clause, it can shift its reference to the matrix subject. (There is still a difference, however. The Slave pronoun can refer to the speaker in the embedded clause, while the judge of the English epistemic modal must be the referent of the matrix subject.) It would certainly be an important result if future work showed that the judge parameter of English epistemic modals behaves in detail like the shiftable indexical of some language, but even if things do not turn out this way, it should be clear that Egan et al.'s

---

[24] A hard-core relativist might not be happy with the formalization in (219), since it does not include the context of evaluation in the semantic entry. Instead, he would want to replace j with an evaluation context c_{eval}.

[25] I omit the convincing arguments that these examples do not involve direct quotes. See the references cited.

argument is based on an excessively narrow assumption about how context dependency works in human language.[26]

Let us now return to the other argument for relativism and see how it fares under this new perspective. MacFarlane believes that (216) is correctly evaluated as false by an eavesdropper who knows that Joe is in Berkeley. One simple point is that I am not convinced of this judgment, and a more careful empirical study is in order here. (See von Fintel and Gillies 2008 for a detailed assessment of the data which has been put forth in favor of relativism.) But, even if we grant MacFarlane's judgment, a different analysis is suggested by Papafragou (2006) and von Fintel and Gillies (2007a, 2007b, 2008). They agree with relativists that the eavesdropper's knowledge might be relevant to the truth conditions of the sentence, but propose that it is relevant through the context of utterance, not the context of evaluation. If this is right, it will be possible to maintain contextualism and reject relativism. This counter-argument begins with the assumption that the proper (contextualist) semantics for epistemic modals can involve more than just the speaker's knowledge. It can involve the knowledge of the speaker and addressee together, or that of the speaker, addressee, and other individuals associated with them. There will often be some amount of vagueness as to exactly whose knowledge is relevant. The eavesdropper might be included among the relevant individuals, and, if he is, what Sally said was indeed false. But, if he is not, what Sally said is true. While in normal conversations there is no issue as to who is among the relevant individuals, the eavesdropper cases have been carefully constructed to make it an issue. No clear principle determines whether the eavesdropper is relevant or not, and therefore Sally's utterance can reasonably be called

---

[26] A very similar argument is made by Moltmann (2005).

I have made this point to philosophers and had them dispute it on the grounds that, even if some languages have shiftable indexicals, English does not. Therefore, the disputation goes, we have no grounds for suggesting a shiftable indexical as part of the semantics of English modals. This response displays a lack of appreciation of the complexity of language in general along with a ill-founded confidence in one's own understanding of English grammar. I also wonder whether it reveals a lack of respect for what we can learn from languages which are very different from familiar Indo-European ones. In the event that this is so, let me point out that Schlenker (2003) argues that French and perhaps English temporal expressions may involve shiftable indexicals. In any case, from the linguists' perspective, if many languages have shiftable indexicals, it's entirely to be expected that they could be found hiding in un-obvious corners of English grammar as well. Please note that, as far as I know, none of the scholars mentioned in the text make this mistake. I am discussing it only because it is a mistake I have heard others make, and wish to go on record with a harsh warning against it.

either true or false. (If we are forced to call it one or the other, unknown pragmatic principles will push us, perhaps reluctantly, to a decision.)

A third point comes from the fact that MacFarlane's position is based on a relatively simple view of the semantics of epistemic modals. In particular, he does not consider the possibility that their semantics should be enriched to express the idea that they are subjective, evidential, or performative in the senses discussed above. To see why this is a real concern, consider the ideas about subjective epistemic modality put forth by Lyons (1977). Lyons thinks that such modals do not contribute to truth conditions, but rather affect the speech act performed by the sentence. If this is the correct way to look at the modal in (216), what Sally said is indeed false: She said (in a reduced-commitment sort of way) that Joe is in Boston right now. Now, I have expressed doubts about Lyons's theory, but the possibility remains that a more sophisticated theory of subjectivity and/or performativity will explain what's going on with eavesdropping examples in a non-relativist way.

In Section 4.2.3, I made a particular proposal concerning the nature of performativity with epistemic modals. I suggested that a sentence of the form *might S* both asserts the modal proposition (roughly $\diamond S$) and adds to the common propositional space the proposition that S. It seems to me that this view allows a simple explanation of MacFarlane's example. He thinks that the eavesdropper will claim that Sally spoke falsely. Suppose this is right. What does it mean to say that Sally spoke falsely? There are two propositions at issue: the modal one, added to the common ground, and the non-modal one, added to the *cps*. The former is true but the latter is false. We can attribute the judgment that Sally spoke falsely to the fact that the non-modal proposition is false.

MacFarlane discusses a proposal similar to this one. He rejects it on the grounds that one can be explicit about what proposition is false:

When you said (supposing you did) that Sally spoke falsely, did you mean that she spoke falsely in saying "Joe might be in Boston," or just that its false that Joe is in Boston? It was the former, right?   (MacFarlane 2006: 6)

However, this response does not work against my version of the idea that epistemic modals are performative. The sentence *Joe might be in Boston* itself involves two distinct propositions, a modal one and a non-modal one. So when the eavesdropper agrees that Sally spoke falsely in saying this sentence, it remains unclear which proposition is being called false. Since there is no natural way to express just the modal proposition asserted by *might S*, I don't think it's possible to reformulate MacFarlane's response in a convincing fashion.

I am suggesting that when the eavesdropper agrees that Sally spoke falsely, she is talking about the non-modal shared proposition, not the modal asserted one. Why would the eavesdropper's judgment focus on the non-modal proposition? It seems to me rather obvious: the eavesdropper has no grounds on which to judge the truth or falsity of the modal proposition, which is after all externally inscrutable (in the sense of Papafragou 2006, see Section 4.2.1). Moreover, I think we can shift the judgments by providing contexts in which it's clear whether the modal or non-modal proposition is relevant:

Third case: You and your friend overhear George and Sally talking in the coffee line. Sally says, "Joe might be in Boston right now." Your friend says, "Oh no, I planned to borrow Joe's car tonight." You say, "That's OK, Sally spoke falsely. I just saw him in Berkeley an hour ago."[27]

Fourth case: You and your friend overhear George and Sally talking in the coffee line. Sally says, "Joe might be in Boston right now." Your friend says, "Sally is deceitful. She knows perfectly well Joe is in Berkeley, but she doesn't want George to meet up with him, since they always go out drinking without her." You say, "Well, no, Sally spoke truthfully. As far as she knows, Joe might be in Boston."

In the third case, what's relevant is whether Joe is in Boston, and here it's natural to judge what Sally said is false. In the fourth case, what's relevant is whether, as far as Sally knows, Joe might be in Boston. Here it's natural to judge that she spoke truthfully. I conclude that there is some evidence that either the modal or the non-modal proposition can be judged for truth or falsity, and that considerations of relevance determine which one we focus on in a particular case. See Portner (2007a) for more discussion of how relevance plays a role in this type of example.

4.3 Issues in the semantics of priority modality

Priority modals have to do with reasons for preferring one situation over another. For example, you can choose to visit Ben or to stay at home. You might prefer the former because Ben is lonely and it's kind to give him company, or because Ben is funny and you enjoy spending time with him, or because Ben is influential and he can help you defeat your enemies. All of these situations can be described with (224):

---

[27] I find MacFarlane's phrasing *Sally spoke falsely* quite awkward here, but I stick with it for the sake of consistency. *What Sally said was false* is more natural.

(224)   You should visit Ben.

In terms of Kratzer's theory we can describe priority modals as having a circumstantial modal base and an ordering source representing reasons for preferring, or assigning priority to, one type of possibility over another.

Priority modals are often called "deontic" by linguists, but as I pointed out earlier, this term can be misleading. In one strict sense, deontic modality has to do with obligations, right and wrong, and other such normative notions. It is useful to have a term for this group, and I have opted to keep "deontic." Therefore, another word was needed for the broader category, and "priority" is appropriate for this job.

4.3.1 Sub-varieties of priority modality

Three of the most common types of priority modality are deontic, bouletic, and teleological. Bouletic modals have to do with someone's desires, and teleological (or goal-oriented) ones with somebody's goals:

(225)   (a)   We must pay the real estate tax. (deontic)
       (b)   Mary should try this new restaurant. (bouletic)
       (c)   John can take the subway. (teleological)

There may be restrictions on the interpretations available with particular modal forms. For example, Kratzer (1981: 61) points out that it is difficult to use German *kann* ("can") with a bouletic interpretation. It is not always easy to classify a particular sentence in terms of these subcategories of priority modality. For example, if Mary's goal is to enjoy a good meal, is (225b) bouletic or teleological? The fact that there is no clear answer to this question means that we should not think of the categories as mutually exclusive or as exhausting the range of meanings. What's essential is that the modal relate to reasons for assigning priority to one alternative over another.

Scholars often make a distinction between two classes of deontics which can be exemplified by the following:

(226)   (a)   At least one son should become a priest.
       (b)   Mary should return the pen she borrowed.
       (c)   You must go home now.

On one reading of (226a), the modal has wide scope: "It is necessary that at least one son becomes a priest." We are not saying that any particular son is under an obligation to become a priest. In contrast, in (226b), Mary is under an obligation to return the pen, and, in (226c), the addressee is under an obligation to leave. We can call the former interpretation the "ought-to-be" reading of the modal, and the latter two an "ought-to-do" reading (Feldman 1986; Brennan 1993). The ought-to-do reading occurs when the subject of the modal is implied to be under an obligation, and otherwise we have the ought-to-be meaning. (Note that (226b) is actually ambiguous. It has an ought-to-be reading, according to which Mary is not under an obligation to return the pen, but rather someone else, for example the speaker, is under an obligation to make sure that she does so. This reading is clearer if Mary is a small child.)

Teleological and bouletic modals show a distinction similar to that between ought-to-be and ought-to-do. Here is a teleological example:

(227) (a) At least two of the students should take the subway.
       (b) Mary should take the subway.

(227b) focuses on Mary's goals, and in this way is like an ought-to-do deontic. (227a), in contrast, does not focus on the goals of the subject. It means something like "In view of our goal of getting all of the students to their destination, at least two of them should take the subway," and in this way is similar to an ought-to-be deontic. However, the teleological and deontic cases differ in that it makes no sense to talk about a goal which is nobody's; in the case of (227a), we are talking about the "our" goals, the goals of the speaker and addressee. In contrast, (226a) need not concern anyone's obligations. It can simply be taken as saying that the situation in which at least one son becomes a priest is (morally or doctrinally) preferable.

Based on the facts in (226)–(227), we may conclude two things: first, that the term "ought-to-do" is misleading, because the distinction is relevant to all classes of priority modals, not just deontics. And, second, that we have to figure out two separate issues:

1. Why does the subject sometimes play a special role in the semantics of priority modals?
2. What is it for an individual to be under an obligation, and how does this idea play a role in the semantics?

We will look at the first question in the next section (4.3.2), and the second in the one after that (4.3.3).

4.3.2 The argument structure of priority modals

Let us first think about how the referent of the subject should affect the meaning of (227b). This sentence says that Mary takes the subway in every circumstantially-accessible world in which her goals are realized. Notice that the set of accessible worlds—those in which her goals are realized—depends on Mary. This suggests that instead of having a simple accessiblity relation (or conversational background), we want to derive the accessibility relation (or conversational background) from the referent of the subject. Working in terms of Kratzer's system, what we need is a functional conversational background, for example:

(228)   The conversational background *teleo* is that function from pairs of an individual *a* and a world *w* to the set of *a*'s goals in *w*.

With generative syntax, it is traditionally claimed that root modals take their subject as an argument (e.g., Ross 1969; Jackendoff 1972; Zubizarreta 1982; Roberts 1985; Brennan 1993). If this is right, it is easy to analyze the subject-sensitivity of examples like (227). We simply need to change the meaning of the modal so that it takes two semantic arguments, a predicate and a subject, and so that it uses its subject argument as one of the arguments of the ordering source:

(229) $[\![\text{ should } \beta]\!]^{c,f,g} = [\lambda x.\{w : \text{for all } u \in \bigcap f(w), \text{ there is a } v \in \bigcap f(w) \text{ such that:}$

    (i) $v \leq_{g(\langle x,w\rangle)} u$, and

    (ii) for all $z \in \bigcap f(w)$: If $z \leq_{g(\langle x,w\rangle)} v$, then $z \in [\![\beta]\!]^{c,f,g}(x)\}]$

(229) describes the meaning of the modal plus the predicate; the subject argument, represented by x, is still unsaturated. (229) correctly predicts that, when the subject is a quantifier, the ordering source can vary with the quantification:

(230)   Both kids should go on a diet.

Perhaps one kid's goal is to look better, the other's to be healthier. Each one's goal implies that he should go on a diet.

The idea that non-epistemic modals take their subject as argument is often strengthened by the hypothesis that root modals are control predicates. Under such a conception, epistemic modals may then be classified as raising predicates. Besides the semantic points we have made already, there are several arguments for the claim that non-epistemic modals take their subject as an argument. For example, epistemic modals freely take expletive subjects; deontic and dynamic modals are sometimes odd with expletive subjects:

(231) (a)   There might be a storm coming.
       (b)   There must be a guard coming. (# deontic)
       (c)   There can be a student swimming in the pool. (# ability)

However, other deontic and dynamic modals are clearly acceptable:

(232) (a)   There must be a guard standing here when I get back.
       (b)   There can sometimes be a lot of lightning around here.

I will discuss the facts concerning dynamic modals in Section 4.4. When it comes to priority modals (specifically, deontic ones), Bhatt (1998), Wurmbrand (1999), and others[28] have used facts like (232) to argue that all modals are raising predicates. That is, they claim that no modals take their subject as an argument. If they are correct, then the analysis sketched in (229) is mistaken. However, while (232a) shows that this particular occurrence of a deontic modal is not a control predicate, it does not show that deontic modals are never control predicates. Perhaps sometimes they are control predicates and sometimes raising predicates. The other arguments offered by Bhatt, Wurmbrand, and others are similar: they show that it's wrong to claim that all priority modals are control predicates, but they do not show that none are. Given the fact that the set of accessible worlds varies with the subject in (230), the simplest conclusion is that the traditional analysis is right about some cases, but not others. That is, some deontic modals are control predicates and some are raising predicates.[29]

4.3.3 Performativity

In trying to understand the nature of the distinction between ought-to-be and ought-to-do deontics, we found that sometimes an individual is implied to be under an obligation. What is it to be under an obligation, and what relevance does this notion have to semantics? It is essential to distinguish two cases before we can answer this question: (i) in some cases the individual who is said to be under an obligation is the referent

---

[28] Barbiers (2001) summarizes much of the recent syntax literature.

[29] Based on a suggestion by Dominique Sportiche, Hacquard (2006: 130) mentions another approach to the sensitivity of the modal to the subject. The idea is that the modal contains a covert pronoun-like argument which may be bound by the subject. Because this proposal is compatible with the claim that priority modals never take their subject as an argument, it is worth pursuing.

of the sentence's subject, while (ii) in others this individual is a participant in the conversation, in particular the addressee. We have dealt with the first case already: when a deontic ordering source is derived from the referent of the subject, as in (228)–(229), the sentence implies that this individual is under an obligation. This treatment extends to other priority modals as well; though we wouldn't say that the goal-oriented (227b) concerns an obligation of Mary, it is focused on Mary in a similar way. This sentence implies that Mary's goals give her a good reason to take the subway.

Now we turn to cases in which the individual who is under an obligation is the addressee or another participant in the conversation. Such cases exemplify performativity. For example, (226c) both asserts that the addressee is under an obligation to leave now and performs an additional speech act of placing the addressee under that obligation. The idea that sentences containing *must* do more than just make a modal statement has been around for a long time (e.g., Leech 1971; Coates 1983; Palmer 1986; Nuyts et al. 2005), and in recent years has received some attention in the theoretical literature as well. In what follows, I will both justify and make more precise the claim that deontic modals are often performative.

An influential early formal analysis of the performativity of deontic expressions is given by Lewis (1979b). Lewis defines a language game involving sentences of the form "!ϕ" (among others); the meaning of this sentence is the same as $\Box\phi$, but when !ϕ is used by a particular player in the game, the "master," to address another player, the "slave," there is an additional effect. The slave is now placed under an obligation to act in accord with ϕ. Lewis's ideas (and the formal model he presents, to be discussed below), are interesting as the first clear expression of performativity in the sense of the term used here. However, it is difficult to determine what the linguistic consquences of Lewis's ideas are, because it is not clear whether "!" is meant to correspond to any particular natural language expressions. Is !ϕ meant to represent an imperative sentence? A sentence of the form *must* ϕ? Is the fact that the slave is placed under an obligation a pragmatic fact, a semantic fact, or some combination of the two? Because these questions are not answered, we should see Lewis's work not as a linguistic analysis, but rather as an exploration of how the notions of obligation and permission fit into communication.

Ninan (2005) develops a system similar to Lewis's, but with a closer attention to linguistic facts. He argues that the deontic modal *must* in a simple root sentence has a two-part meaning: first, it is an ordinary modal operator, so that the sentence expresses a proposition involving quantification over possible worlds. And, second, it acts like an imperative, placing an obligation on the addressee.[30] This analysis can explain facts like the following:

(233) (a) Mary must leave now.
       (b) # Mary must leave now, but I know she won't.
       (c) Mary must have left. (# deontic)

According to Ninan, when I utter (233a), I both assert that Mary leaves in all deontically accessible worlds and place a requirement on the addressee to ensure that Mary leaves. According to this theory, (233b) is unacceptable because I cannot place a requirement on someone using *must* while at the same time presupposing that it will not be met. (This restriction is a stipulation, but it is similar to one which applies to imperatives: *Leave!*—*#even though I know you won't.*) (233c) is bad on the deontic reading because one cannot impose an obligation concerning the past.[31]

While *must* is performative, according to Ninan, *should* is not. Thus, sentences like (233b–c) containing *should* are fine:

(234) (a) Mary should leave now.
       (b) Mary should leave now, but I know she won't.
       (c) Mary should have left (but she didn't).

Of course, it's possible for a speaker to use a sentence with *should* to place a requirement on the addressee. For example, if the officer says to the soldier (234a), he may well take it as an order to escort Mary from the room. However, Ninan would analyze this as a Gricean implicature, rather than a part of the conventional meaning of *should*. Because it is

---

[30] Interestingly, Ninan's analysis of *must* is very much like the one Lewis (1972) gives of imperatives. Lewis proposes that an imperative both has the truth conditions of *I command you to* . . . and performs a directive speech act. Portner (2007b) has further discussion of the relation between modals and imperatives.

[31] (233c) is acceptable on an interpretation in which the leaving is in the future:

(i) Mary must have left (by the time I get back).

Ninan actually predicts that (i) is bad as well, because he thinks that the perfect form in modal sentences functions as a past tense operator. However, as pointed out by Portner (2007b), the perfect can have its ordinary aspectual meaning in modal sentences like (i). Given that the perfect is aspectual, and allows the event of leaving to be in the future, the grammaticality of (i) is expected under Ninan's analysis of *must*.

just an implicature, this imperative-like meaning can be cancelled, as in (234b–c).

Ninan also points out that some embedded occurrences of *must* are not performative. (235a–b) do not place a requirement on the addressee, while (c) does:

(235) (a) John thinks that Mary must leave now.
       (b) If Mary must leave now, I will leave too.
       (c) Mary must leave now, and you can too if you want.

If we accept that *must* has an imperative-like meaning, in addition to an ordinary modal meaning, we next must ask what the nature of this meaning is. The work of Lewis (1979b) has had a tremendous influence on semanticists who have aimed to answer this question. Recall that Lewis develops his ideas about the performative component of meaning within the context of a metaphorical language game involving a "master" and a "slave." (There's also a "kibitzer"). A sentence of the form !ϕ has a special role in the language game: it is a modal statement, and—when the master addresses the slave—places the slave under an obligation. Here are the relevant semantic definitions:

(236) **Semantics of Lewis (1979b)**

       (a) A function f assigns any pair of a time and a world $\langle t, w \rangle$ a set of worlds, the SPHERE OF ACCESSIBILITY.
       (b) A function g assigns any pair of a time and a world $\langle t, w \rangle$ a set of worlds, the SPHERE OF PERMISSIBILITY.
       (c) For any time t and world w, $[\![\ ]\!]^{\langle t, w \rangle}$ assigns to any sentence ϕ a truth value 0 or 1.
       (d) For any sentence α of the form !ϕ, $[\![\alpha]\!]^{\langle t, w \rangle} = 1$ iff $[\![\phi]\!]^{\langle t, v \rangle} = 1$, for every $v \in f(\langle t, w \rangle) \cap g(\langle t, w \rangle)$.
            (i.e., ϕ is true in every world which is both accessible and permissible.)
       (e) For any sentence α of the form ¡ϕ, $[\![\alpha]\!]^{\langle t, w \rangle} = 1$ iff $[\![\phi]\!]^{\langle t, v \rangle} = 1$, for some $v \in f(\langle t, w \rangle) \cap g(\langle t, w \rangle)$.
            (i.e., ϕ is true in some world which is both accessible and permissible.)

Here, $f(\langle t, w \rangle)$ is the set of worlds which are compatible with the history of w up until time t. They are exactly like w up until t, and differ only in the future after t; that is, f is a historical accessibility relation.

The set of worlds which are permissible, $g(\langle t, w \rangle)$, provides the range of options available to the slave. The slave must act so as to keep the world within this set. Formally, g is also an accessibility relation; one can think of its function as similar to a deontic ordering source, in that it brings in information concerning what is permissible. But it is crucially different from an ordering source in that it does not provide an ordering; like any accessibility relation, it just gives a set of worlds.

Points (236d–e) provide the semantics for operators ! and ¡. The former is a $\square$ and the latter a $\diamondsuit$. The performativity of these operators is given by the following principles:

(237)   **Performativity in Lewis (1979b)**

    (a)   At any time t, the slave must act so as to keep the actual world within the sphere of permissibility $g(\langle t, w_{actual} \rangle)$

    (b)   For any sentence a of the form $!\phi$ or $¡\phi$, if the master says a to the slave, the sphere of permissibility changes so as to make a true.

        (i)   If a of the form $!\phi$ and $f(\langle t, w \rangle) \cap g(\langle t, w \rangle)$ contains worlds in which ϕ is false, g changes so that $f(\langle t, w \rangle) \cap g(\langle t, w \rangle)$ does not contain any worlds in which ϕ is false.

        (ii)   If a of the form $¡\phi$ and $f(\langle t, w \rangle) \cap g(\langle t, w \rangle)$ does not contain any worlds in which ϕ is true, g changes so that $f(\langle t, w \rangle) \cap g(\langle t, w \rangle)$ contains some worlds in which ϕ is true.

(237a) explains the practical importance of the sphere of permissibility. It constrains the actions of the slave. If the sphere only contains worlds in which the slave carries rocks all day, the slave will be in trouble if he does not carry rocks all day. (237b) says how the sphere of permissibility can be changed by the master. Part (i) says that, when it is used by the master to the slave, $!\phi$ restricts the sphere of permissibility; in effect, $!\phi$ is a command. If the master says *!Slave carries rocks all day*, the slave will be in trouble if he does not. Point (ii) says that $¡\phi$ expands the sphere of permissibility; $¡\phi$ gives permission. If the master says *¡Slave takes tomorrow off*, the slave will not be in trouble if he does not carry rocks (provided he behaves properly otherwise).

Sentences of the form $!\phi$ are not exactly like any natural language expressions. Their performative meaning is similar to that of an imperative, and indeed Lewis's analysis has influenced semantic theories of imperatives. However, !ϕ can be used to make a statement; for example, if the master says *!Slave carries rocks all day* to the kibitzer, the meaning is similar to "The slave has to carry rocks all day." In this respect, it differs from imperatives, since they cannot be used to make a statement. Actually, !ϕ is more similar to a sentence with *must*, but it differs in that the performative meaning only arises when it is used by the master to the slave; in contrast, if Ninan is right, an unembedded sentence with *must* is always performative, placing a requirement on the addressee.

Neither of the statements given in (237b) is a complete definition. Much of Lewis's paper is devoted to the fact that it is easy to extend (i) to a complete definition of commanding, but it is difficult to see how to extend (ii) to a complete definition of permission. Let's start with (i): this says that when the master commands !ϕ, the sphere of permissibility must shrink so as to include only worlds in which ϕ is true. How much does it shrink? Lewis proposes that it shrinks only as much as is needed. In particular:

(238)   If !ϕ is addressed by the master to the slave at time t, $g(\langle t, w\rangle)$
         $= g(\langle t_{-1}, w\rangle) \cap [\![\, \phi \,]\!]^{\langle t, w\rangle}$, for any world w.

(t_{-1} is the time just before t.) Turning to (ii), one might think that ¡ϕ requires a principle just like this one, except that it uses union to expand the sphere of permissibility rather than intersection to shrink it:

(239)   If ¡ϕ is addressed by the master to the slave at time t, $g(\langle t, w\rangle)$
         $= g(\langle t_{-1}, w\rangle) \cup [\![\, \phi \,]\!]^{\langle t, w\rangle}$, for any world w.

But this does not work. Under this definition *¡Slave takes tomorrow off* would make permissible any world in which the slave takes tomorrow off. This includes worlds in which the slave takes takes tomorrow off and kills the master; clearly permission in natural language does not work this way. Therefore this definition is not helpful for gaining a better understanding of how permission works in human communication. Lewis (1979b) contains useful discussion of what it would take to develop a better account than (239).

Ninan develops a more linguistically realistic account of the performativity of deontic modals. In the first place, he does not define where performativity arises with an arbitrary definition like Lewis's, but rather describes (in a sketchy way) its distribution. As pointed out above, his main observations are that it occurs with *must* but not *should*, and that it disappears in certain embedded contexts. In addition, Ninan bases his analysis of the performative component of meaning on a linguistically realistic theory of imperative semantics, that of Portner (2004). In particular, he assumes that each participant in the conversation is associated with a To-do List, and that the function of an imperative is to add the proposition it expresses to the addressee's To-do List.[32]

It is simple to use the To-do List to define a sphere of permissibility: if $T(x)$ is x's To-do List, x must act so as to keep the actual world within the set $\bigcap T(x)$. In terms of this idea, we can understand patterns like the following:

(240) (a) Sit down! # You're not going to sit down.
       (b) # Have sat down before now!

(240a) is unacceptable because one cannot impose a requirement on someone if one assumes they will not carry it out; more precisely, $T(x)$ must remain compatible with the common ground, so one cannot add a proposition p to $T(x)$ if p is incompatible with the common ground. (240b) is bad because one cannot impose a requirement pertaining to the past; more generally, one cannot at time t add p to $T(x)$ if no actions one can take after t can have a bearing on whether p is true.[33]

The facts in (240) are completely parallel to the those pertaining to *must* in (233). Ninan proposes that a root utterance of *must* ϕ adds ϕ to the addressee's To-do List, and as a result (233b) is unacceptable just as (240a) is, and (233c) is unacceptable just as (240b) is. These ideas can be formalized in a dynamic framework as follows:

(241) **Update potential of *must* (deontic):** For any sentence ϕ of the form *must* ψ, the update potential of ϕ used in context c with modal base f and ordering source g, $[[\,\phi\,]]^{c,f,g}$, is defined as follows:

$\langle cg, T \rangle [[\,\phi\,]]^{c,f,g} = \langle cg', T' \rangle$, where

(i) $cg' = cg \cup \{[[\,\phi\,]]^{c,f,g}\}$, and
(ii) T' is just like T except that
      $T'(addressee) = T(addressee) \cup \{[[\,\psi\,]]^{c,f,g}\}$

[32] Portner (2004) treats a To-do List as a set of properties, not propositions, but my reasons for doing so are not important to Ninan. Therefore he simplifies the treatment by working with propositions.

[33] Ninan does not give formal versions of the principles we need to explain the facts in (240), and here I just sketch the principles he seems to have in mind.

Let's use (233a) as an example. Point (i) is the expected meaning in possible worlds semantics; the sentence conventionally asserts that Mary leaves in all accessible worlds. Point (ii) is the performative meaning; it adds the proposition that Mary leaves to the addressee's To-do List.

While this way of using the To-do List would suffice for Ninan's purposes, it has the same problems as Lewis's sphere of permissibility. In particular, it does not allow an adequate definition of the function of permission sentences. But Portner (2004, 2007b) actually proposes that the To-do List is not used in the simple way outlined above to generate a sphere of permissibility. Rather, a To-do List has a function analogous to that of an ordering source: x's To-do List $T(x)$ establishes an ordering of the set of worlds compatible with the common ground, and x is is committed to taking actions which tend to make the actual world as highly-ranked as possible in terms of this ordering, given the propositions accepted in the common ground. In other words, the To-do List brings ordering semantics into the analysis of directive speech acts. The use of ordering may allow a solution to the problems associated with permission, but further discussion will have to wait another occasion (in particular, Portner, forthcoming; see also van Rooij (2000)).

If Ninan is right, imperatives and deontic modals are closely related in their pragmatics. There are links between imperatives and other priority modals as well. Portner (2007b) finds that the range of interpretations available to imperatives matches those found with priority modals generally. For example:

(242)   (a)   Sit down right now! (order)
        (b)   Have a piece of fruit! (invitation)
        (c)   Talk to your advisor more often! (advice/suggestion)

The imperative in (242a) gives an order, and as such has an obvious relation to a deontic interpretation. That in (242b) invites the addressee to do something because it would provide enjoyment; this is parallel to a bouletic interpretation. And, finally, (242c) makes a suggestion concerning what the addressee ought to do in order to reach his goals; thus, it can be related to a teleological interpretation.

This parallelism between imperatives and priority modals confirms Ninan's idea that the semantics of deontic modals should be linked to that of imperatives, but it makes clear that there is more to the link than performativity. The range of interpretations seen with priority modals is explained, in Kratzer's system, in terms of the variety of ordering sources available to these modals. The meaning of an imperative is modeled by adding the proposition (or property) it expresses to the To-do List. Therefore, if we want to give a precise model of the connection between priority modals and imperatives, the To-do List must have a way of representing deontic-like (order), bouletic-like (invitation), and teleological-like (advice) meanings for imperatives. See Portner (2007b) for further discussion and a precise theory of the relationship between ordering sources and To-do Lists.

4.4 Issues in the semantics of dynamic modality

Under the umbrella of dynamic modality, I include two types: volitional and quantificational modality. Quantificational modals have received very little attention in the literature, and this fact has led to confusion about the nature of dynamic modality. Linguists typically exemplify the concept of dynamic modality with modals of ability, disposition, and the like. These are the core examples of dynamic modality. However, as we will see below, they have not drawn a clear distinction between volitional and quantificational modals, and as a result non-volitional, quantificational modals are sometimes considered dynamic as well. These practices have left us with a problem of terminology, and I have chosen to resolve it by using "dynamic" for the broad class which includes everything anyone has wanted to call dynamic, plus quantificational modals. I employ "volitional" for the narrower category which includes the most well-known types of dynamic modality, in particular modals of ability.

4.4.1 Volitional modality

Volitional modality includes at least the meanings of ability and opportunity, as illustrated in (243a–b). Many linguists also claim that *will* can express volition, and if this is so (a relevant example would be (243c)), it should be included as well.

(243) (a) John can swim.
       (b) Mary can see the ocean.
       (c) He will come, if you ask him. (Palmer 1986: 103)

In (243a), *can* indicates the subject's intrinsic ability, while (243b) has to do with the situation in which the subject finds herself. Example (243c) suggests that the subject has the disposition or willingness to come.

There probably exist other kinds of modality which should fall into the same category as those in (243), but which are not naturally called "volitional." For example, H. Davis et al. (2007) discuss the "out of control" form *ka-...-a* in St'át'imcets. They reduce its various intuitive meanings to two basic interpretations, ability and involuntary action. Obviously "volitional" is a poor term for the latter. Nevertheless, their analysis implies that it is a type of non-priority, non-epistemic element closely related to ability modals. If further research shows this type of modal meaning to be common, it will be necessary to rethink our terminological choices.[34]

Some descriptive linguists exclude volitional modals from the field of true modality on the grounds that they do not display subjectivity or performativity (e.g., Palmer 1986). However, they are modals in the most clear semantic sense—they allow us to make statements which depend on non-actual situations—and in many languages should be grouped with epistemic and priority modals on the basis of morphosyntactic properties as well. Therefore, the decision to exclude them from the field of true modals on the grounds that they do not have vaguely defined properties of subjectivity and performativity does not seem well justified.

Problems in defining volitional modality

While it is fairly easy to identify core examples of volitional modality, matters become much less clear as we consider other cases:

(244) (a) *John can fear his teacher.
       (b) A student can fear his teachers.
       (c) Mary can understand French.
       (d) It can snow on Mt Jade.

The fact that (244a) is ungrammatical suggests that volitional *can* places some sort of requirement on its subject. It has been proposed that the subject must be an agent or cause of the event described by

---

[34] H. Davis et al. (2007) also make the proposal that modals in St'át'imcets lexically code for the types of conversational backgrounds they are compatible with, but allow either necessity or possibility as their modal force. In this, they appear to differ from modals in English, which code more for force than for type of background, and so the paper offers an interesting account of cross-linguistic variation. However, in another paper, Rullmann et al. (2007), they appear to adopt a different analysis, mentioned in Section 4.1.3. For this reason, I don't discuss the theoretical aspects of H. Davis et al. (2007) further.

the sentence (Hackl 1998). However, if that is correct, why is (244b) grammatical? Examples (244c–d) also pose serious challenges for the idea that volitional *can* places a restriction on the thematic role of its subject.[35]

One problem that has afflicted previous analyses of volitional modality is the fact that it is not clear how we should draw the line between volitional and quantificational modals. Since most discussions of modality ignore or set aside quantificational modals, they do not face this issue directly. However, once we do face it, it becomes apparent that some examples which have been difficult to analyze as volitional modals are actually quantificational modals instead. In particular, the modals in (244b–d) might well be quantificational. (I think that (244a) also has a difficult-to-perceive quantificational reading.) This is most clear in the case of (244d)—Brennan (1993) lists examples very similar to this one as quantificational—but (244b–c) might be too. Let me explain informally how they can be thought of as quantificational; in Section 4.4.2, we will delve into the semantics of quantificational modals in more detail.

The contrast between (244a) and (244b) can be explained in terms of two assumptions. The first is that an ability meaning for *can* is ruled out here because the subject is not an agent or a cause. The second is that *can* also functions as a quantificational modal with a meaning similar to an adverb of quantification. Recall that, according to Lewis's (1975) original theory of adverbs of quantification (and its direct descendants Discourse Representation Theory, Kamp 1981, and File Change Semantics, Heim 1982), such an adverb is able to bind the variables introduced by indefinites. More specifically, Lewis proposes that a sentence with a quantificational adverb assumes a tripartite structure at logical form: the quantifier, its restrictor, and its scope. In the case of (244b), the subject is the restrictor and the rest of the sentence is the scope:

[35] Hackl suggests that sentences like (244c) have an agentive meaning ("Mary has the ability to achieve understanding of an utterance in French"), but this paraphrase assigns Mary a more active role than (244c). Moreover, the use of *can* here seems much more natural than other words which would trigger an agentive interpretation, like *intentionally*:

(i)   ??Mary intentionally understood French.

Example (244d) creates an even more severe problem. Hackl suggests that this sentence describes an act of causation without representing the causer, and that the result is a dispositional meaning. However, he does not explain how the dispositional reading arises or how it is related to the ability reading.

(245)

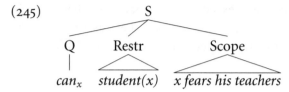

$$can_x \quad student(x) \quad x\ fears\ his\ teachers$$

Given this logical form, *can* binds the variable introduced by *a student*, and therefore quantifies over students. The sentence then receives the more-or-less correct meaning "Some students fear their teachers." In contrast, (244a) is ungrammatical because there is no indefinite which can provide a variable for *can* to bind (and the ability reading is ruled out because the subject does not have the right kind of thematic role).

Next let us look at (244d). Assuming that this sentence contains a quantificational modal, what constitutes its restrictor? Though there is no indefinite noun phrase in the sentence, another possibility suggests itself: the PP. The PP describes locations or situations, and we can think of *can* as quantifying over these.

(246)

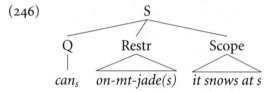

$$can_s \quad on\text{-}mt\text{-}jade(s) \quad it\ snows\ at\ s$$

This logical form looks promising, but several questions must be answered before we can understand exactly what meaning it represents. First of all, what is a situation or location? And, second, how are these things linked to the main predicate in the clause *it snows*? In other words, what do we mean by *it snows at s*? We will return to this issue in Section 4.4.2.

Finally, we turn to (244c). Before looking at it directly, we will first examine an example involving an adverb of quantification instead of a modal. Along the lines of Kratzer (1995), we can think of the adverb in (247) as quantifying over locations or situations:

(247) (a) When you speak clearly, Mary always understands French
       (b)

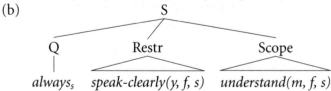

$$always_s \quad speak\text{-}clearly(y, f, s) \quad understand(m, f, s)$$

The intended paraphrase for this structure is "In all situations *s* in which you speak French clearly to her, Mary understands."

Example (244c) is like (247) except for two facts: we have *can* in place of *always* and the restrictor is unexpressed. Let's assume that the restrictor is filled in by context, here represented by the variable C:

(248)

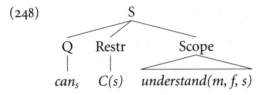

This tree represents a meaning like "In some contextually relevant situations, Mary understands French." I hope that at this point it is at least clear what I mean by the suggestion that the examples in (244b–d) are quantificational. We will examine the semantics of quantificational modals in more detail in Section 4.4.2.

If we provisionally assume that the examples in (244b–d) involve quantificational modals, it is possible to maintain the idea that the subject of a volitional modal in English is always an agent or cause. At the very least it is always volitional, i.e. a sentient individual who is willfully involved in the event or events described by the main predicate. (We saw above that other languages may well have a broader class of non-quantificational dynamic modals, but we set this aside for now, as we try to understand the issues which have been confusing in the English literature.) This is why I call the class of non-quantificational dynamic modals "volitional." The next question is how to enforce the requirement that the subject be volitional. As pointed out in Section 4.3, the traditional explanation is that non-epistemic modals take their subject as an argument, while epistemic modals do not (Ross 1969; Perlmutter 1971; Jackendoff 1972; Zubizarreta 1982; Roberts 1985; Brennan 1993, among others). This difference in argument structure is often related to the difference between control and raising verbs, leading to the description of non-epistemic modals as control predicates and epistemic modals as raising predicates. In recent years, several authors have argued against this perspective (e.g., Bhatt 1998; Wurmbrand 1999; Barbiers 2001). However, these argument relate only to priority modals, in particular deontic modals.[36] As far as I know, the only grounds on

---

[36] As discussed in Section 4.3, all these arguments show is that not all deontic modals are control predicates. Wurmbrand, for example, assumes that all root modals have the same structure, and then argues against the hypothesis that they are control predicates. She does this by showing showing some deontic modals which behave as raising predicates. However, it could be that some deontic modals are raising predicates, while others, and dynamic modals as well, are control predicates. This is the perspective I adopt.

which to challenge the idea that volitional modals take their subject as argument arise from examples like those in (244), and, as we have seen, these may well be quantificational rather than volitional. Therefore I will assume that volitional modals take a subject argument. They are in fact control predicates and thus are able to place an appropriate semantic restriction on their subject.

The modal analysis of ability

Kenny (1975, 1976) has argued that ability cannot be analyzed in terms of a possibility operator within modal logic (see also Horty 2001 for discussion). He considers the case of the dart player who is skilled enough to hit the board, but not skilled enough to hit any particular area of the board. In that case, we'd say that (249a) is true, but neither (249b) nor (249c):

(249)   (a) He can hit the board.
             He has the ability to hit the board.
       (b) He can hit the top half of the board.
             He has the ability to hit the top half.
       (c) He can hit the bottom half of the board.
             He has the ability to hit the bottom half.

These facts pose a problem, if *can* represents a $\Diamond$ within modal logic, because $\Diamond p$ (= 249a) entails $\Diamond q \vee \Diamond r$ (= (249b) or (249c)), where $p \leftrightarrow (q \vee r)$. Fundamentally there is a problem for the idea that (249b) means "there is a world compatible with his abilities where he hits the top half" because this analysis allows that his hitting the top half is only luck. In fact, the English sentence with *can* (or *have the ability to*) requires that he have control over the outcome.

An argument similar to Kenny's has been made by Hackl (1998). Hackl specifically says that ability modals pose a problem for Kratzer's theory of modality. We should check whether this really is so by making explicit how that theory analyzes volitional modals. In all of the cases in (243), stable intrinsic facts concerning the subject are relevant. For example, in (243a) the modal base should include propositions expressing John's physical attributes, learned skills, and so forth. In other examples, temporary or extrinsic facts are also relevant; thus, in (243b) we have to consider Mary's location and the weather conditions. Example (243c) focuses in particular on the subject's desires and plans. The easiest way to represent these differences is to say that, in all cases, the modal base includes propositions expressing the subject's stable, intrinsic qualities; in (243b), the ordering source brings temporary and extrinsic facts into the picture and in (243c) it introduces the subject's (perhaps temporary) plans and desires. Other ordering sources are possible as well; Kratzer (1981: 60–1) points out that German *kann* ("can") allows a "normal standards" ordering source (her (51)):

(250) Ich kann das nicht aushalten.
      I   can this not   bear
      'I cannot bear this.'

As with other types of modality, languages differ in the precise range of meanings that the volitional modals can express. For example, in its ability use Mandarin *hui* can only be used to express what an individual can do in light of his stable, intrinsic qualities. Returning to the challenge posed by (249), if we assume that *can* expresses (human) possibility within Kratzer's system, the problem arises in exactly the same way that it does in modal logic.

Several similar analyses of ability modals have been given with the aim of solving the problem brought up by Kenny, e.g., Brown (1988), Horty (2001), and Hackl (1998). All of these approaches are alike in combining some sort of existential quantification, corresponding to the idea that the agent chooses an action, and some sort of universal quantification, corresponding to the idea that the action guarantees a certain outcome, like hitting the board. The approaches differ significantly in how they work out this general intuition. In Brown's theory, the two "layers" are combined into a single operator. In Horty's, they are the result of combining a standard possibility modal (a $\Diamond$) with a special operator (called a "see to it that" or *stit* operator) defined in terms of the branching time model (on branching time, see Sections 2.3.3 and 5.1.1). Hackl agrees that there are two operators, and that the first is an ordinary modal; he proposes that the second operator is voice,[37] heading a VoiceP projection and introducing the notion of agency or cause into the semantics. He does not provide a detailed semantics for voice, however.

Hackl follows Kratzer herself in the hypothesis that the modal force of ability modals is ordinary ("human") possibility. However, we should recall from the discussion in Section 3.1 that Kratzer actually defines two other types of possibility operators, weak possibility and good possibility. In fact, if ability modals represent the force of good

---

[37] Voice is the name of the linguistic category which distinguishes active and passive sentences, among other types. The idea of a voice operator and a VoiceP is that a particular grammatical element is responsible for determining which voice a particular sentence is.

possibility, Kenny's problem does not arise. A proposition p is a good possibility (with respect to a modal base and an ordering source) iff there is some world u compatible with the modal base, and for all better ranked worlds v (according to the ordering source), p is true in v. Consider that (249a) could be true because there is some u compatible with the modal base where there is an infinite sequence of better worlds as follows:

$$(251) \quad \ldots v_{top} \leq v_{bottom} \leq v_{top} \leq v_{bottom} \leq u_{top}$$

(Here v_{top}/v_{bottom} means that v is a world in which he hits the top/bottom half of the board.) The important thing to notice here is that worlds in which he hits the top and worlds in which he hits the bottom alternate as we move down an infinite sequence of ever better worlds. We might think of this as representing the fact that small changes in the situation (e.g., the amount of breeze, the amount of beer he has drunk, and so forth) determine which half he hits, and crucially that his skill does not control where the dart strikes.[38] This means that (249a) can be true, while neither (249b) nor (249c) is. The reason that good possibility escapes the problem is that it incorporates two levels of modal quantification—"**there is** a world u such that **for all** better worlds v"— and this makes it like the solutions of Brown and Horty.

The actuality entailment

Bhatt (1999) demonstrated an important property of ability modals: in languages that make a clear distinction between perfective and imperfective aspect, perfective sentences containing these modals entail that an event of the kind described by the sentence (minus the modal) actually occurred (example from Bhatt 1999, (321)).

(252) (a) Yusuf havaii-jahaaz uṛaa sak-taa   hai/thaa
           Yusuf air-ship   fly can-impfv be.pres/be.pst
           (lekin vo havaii-jahaaz nahĩĩ uṛaa-taa hai/thaa).
           but   he air-ship   neg fly-impfv be.pres/be.pst
                                                              (Hindi)
           'Yusuf is/was able to fly airplanes (but he didn't fly airplanes).'

---

[38] A more realistic model of this situation wouldn't have such a strict alternating of top and bottom worlds. All that's crucial is that, for each top world, there's a better bottom world, and vice versa. There can be groups of top and bottom worlds which are tied in the order, and clusters of top or bottom worlds.

    (b)   Yusuf havaii-jahaaz uṛaa sak-aa   (lekin us-ne
           Yusuf air-ship   fly   can-pfv but   he
           havaii-jahaaz nahĩĩ uṛaa-yaa.
           air-ship   neg   fly-pfv
           'Yusuf is/was able to fly airplanes (#but he didn't fly air-
           planes).'

This type of entailment is known as the ACTUALITY ENTAILMENT.

    Recently Hacquard (2006) has shown that the actuality entailment is not limited to ability modals, but also occurs with some priority modals. For example (Hacquard 2006, (45)):

(253) (a)   Lydia a   pu   aller   chez sa   tante
             Lydia has   could-pfv go-inf to   her aunt
             (selon   les   ordres de son père). (French)
             according-to the   orders of   her   father
             'As a result of her father's orders, Lydia was allowed to go to her aunt's house (and in fact went).'

       (b)   Lydia a   dû   faire   la   vaisselle (selon
             Lydia has   must-pfv do-inf the dishes   according-to
             les ordres de son père).
             the   orders of   her   father
             'As a result of her father's orders, Lydia had to do the dishes (and in fact did them).'

Epistemic modals don't show the actuality entailment, and while imperative-like (performative) deontics are not very good with perfective morphology, to the extent that they are acceptable, they also lack the entailment (her (40a) and (50)[39]):

(254) (a)   (Selon   la   voyante,)   Bingley a
             (According.to the fortune-teller) Bingley has
             pu   aimer   Jane.
             could-pfv love-ing Jane
             'According to the fortune teller, Bingley might love Jane.'

---

[39] Hacquard's (50) is slightly more complex than the example given here. It is judged virtually unacceptable on the interpretation where the speaker aims to place a requirement on the addressee, i.e. on an imperative-like reading, and, to the extent that it is acceptable, it does not show the actuality entailment.

(b)   ??Kitty a   été   censée/supposée faire   ses
      Kitty   has   was-pfv supposed   do-inf her
      devoirs.
      homework
      'Kitty was supposed to do her homework.'

I will first review Bhatt's and Hacquard's analyses of these phenomena, and then make some further observations.

Bhatt's theory

Bhatt argues that the morpheme *sak* which is glossed as *can* in (252) is actually not a modal at all, but rather an implicative predicate with a meaning similar to *manage*. He uses *ABLE* as a gloss for the implicative *sak*. (252a) means something like "Yusuf managed to fly airplanes (but he didn't)," which is plainly contradictory. He goes on to propose that when the verb form is imperfective, a higher operator is present and removes the actuality entailment. In particular, imperfective aspect contributes a generic operator, leading to logical forms like the following:

(255)   *Past(Gen(ABLE(*Yusuf flies airplanes*)))*

The generic operator *Gen* is assumed to be intensional, so that in *Gen(φ)* does not entail *φ*. More particularly, he assumes that *Gen* is precisely the operator whose meaning we can detect more easily in sentences like (256):

(256)   This machines crushes oranges. (Bhatt 1999, (342a))

Notice that (256) does not entail that the machine ever has or will crush oranges; it just means that it is built in such a way that it can crush them. The fact that it does not entail that the machine will actually crush any oranges is by hypothesis due to the presence of *Gen* in the sentence. Returning to (252a), we clearly perceive an ability meaning: Yusuf has the ability to fly airplanes. Bhatt claims that this meaning is the result of having *GEN* in the sentence.

It is interesting that Bhatt's analysis of ability sentences is in some ways similar to the proposals of Horty (2001) and Hackl (1998) for how to deal with Kenny's problem, discussed above with respect to (249). All of these authors propose that sentences expressing ability involve two operators, one which introduces existential quantification over worlds (*GEN* for Bhatt, $\Diamond$ for Horty and Hackl) and one which introduces some sort of agentive meaning (*ABLE* for Bhatt, *stit* for Horty, and voice for Hackl). Perhaps we could make the semantics of Bhatt's proposal more precise by replacing *GEN* and *ABLE* with the corresponding components of Horty's analysis, $\Diamond$ and *stit*. We must keep in mind, though, that it is essential to Bhatt's account of the actuality entailment that the form *sek* which appears overtly in (252) corresponds to the lower one of these, namely *ABLE*. That is, according to Bhatt, what appears to be a modal in ability sentences actually is not a modal; if a modal is present, it is null. This feature leads Hacquard to make several arguments against Bhatt's theory.

Hacquard criticizes Bhatt's analysis in three ways. First, she points out that it leaves unexplained why the implicative verb meaning *ABLE* can typically also function as a modal. For example, the French *pouvoir* ("may," "can," "could") and *devoir* ("must") give rise to the actuality entailment on some of their uses, but can also serve as epistemic and deontic modals. According to Bhatt, this difference amounts to a lexical ambiguity between an implicative verb and a modal. Second, she points out that it is difficult to extend Bhatt's analysis to priority modals like (253a–b). For example, suppose that *a pu* (form of *pouvoir*, "can") actually means *ABLE* when it gives rise to the actuality entailment; what is the difference between (253a), where it seems to have a deontic meaning, and other cases where it seems to express ability? Bhatt would have to propose multiple implicative verbs, one designed to capture the ability flavor of (252), another designed to capture the deontic flavor of (253). And, finally, it is difficult to extend Bhatt's analysis to necessity modals like *a dû* (form of *devoir*, "must") in (253b). This sentence entails that there was no other way Lydia could follow her father's orders other than by doing the dishes, in contrast to the (a) example which only says that going to her aunt's house was compatible with her father's orders. It is difficult to see how to capture this difference on Bhatt's theory. Rather, it strongly suggests that *pouvoir* is a weak (possibility) modal and *devoir* is a strong (necessity) modal.

Hacquard's theory

Hacquard's analysis is similar to Bhatt's in its overall strategy. Like Bhatt, she provides an explanation for why perfective sentences have the actuality entailment, and argues that the reason for the absence of the entailment when the morphology is imperfective is that there is another operator in the structure. Her proposal differs from Bhatt's in that she assigns the modal forms in these sentences the familiar modal meanings

from possible worlds semantics. The key to deriving the actuality entailment with perfectives is the idea that aspect binds a variable across the
modal. She derives such a configuration by having the perfective aspect
undergo movement from from a position close to the verb to a position
higher than those modals which give rise to the entailment (but below
others). In particular, she says that aspect starts out as an argument
of the verb and raises to a position higher than ability modals, goal-
oriented modals, and ought-to-do deontics; this position is said to be
lower than epistemics and ought-to-be deontics, explaining why these
modals do not give rise to the actuality entailment. (In my terminology,
the relevant position would be below epistemics and imperative-like
(performative) priority modals, and above other modals.)

   After the aspectual operator moves, it binds its trace, which represents an event argument of the verb:

(257) (a) Jane could-pfv run. (Hacquard 2006, (75); tree and for-
             mula are simplified)

      (b)

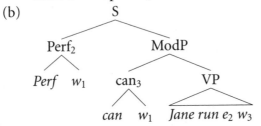

      (c) $\exists e_2[e_2$ in $w_1 \wedge e_2 \subseteq t \wedge t < \text{now} \wedge$
             $\exists w_3[R(w_1, w_3) \wedge e_2$ is an event of Jane running in $w_3]]$

In this simplified tree, one can think of w_1 as the actual world. The
modal *can* takes w_1 as an argument, meaning that it quantifies over
worlds accessible from w_1. The variable w_3 represents those accessible
worlds, i.e. the worlds in which Jane realizes her abilities. The perfective
operator also takes a world argument, and it binds an event variable e_2
within the VP; the e_2 represents the events of Jane running.

   The formula in (257c) expresses the meaning within Hacquard's theory. The first line of the formula corresponds to the perfective operator. It
says that e_2 exists in the actual world and is a perfective event. The
second line gives the semantics of the ModP. Here we have an ordinary
possibility modal quantifying over worlds accessible by the relation R.
The key point to note is that the variable e_2, introduced by the perfective operator, occurs again. The portion of the formula corresponding

to the VP says that this perfective event e_2 is an event of Jane running in w_3.

Note that e_2 is asserted to exists in both w_1 and w_3. This is essentially where the actuality entailment comes from, according to this analysis. In those worlds (w_3) in which Jane realizes her abilities, it is a running event; moreover, it is an actual event. This almost says that Jane actually ran, but not quite. Nothing explicitly says that this event is a running event in the actual world. Things may have different properties in different worlds, and this goes for events as much as for anything else. Hacquard gives the following example, from Kai von Fintel (p.c.):

(258)   Bill mistakenly thought that Mary's wedding was a funeral.

This sentence seems to imply that there are possible worlds (those which are described by Mary's beliefs) in which the wedding was a funeral. If we accept this way of thinking about (258), we cannot assume, simply because e_2 is an event of Jane running in some other world (w_3), that it is also an event of Jane running in the actual world (w_1).

In order to derive the actuality entailment, Hacquard must find a way to ensure that an event e_2 which is an event of Jane running in the accessible worlds w_3 is also an event of Jane running in the actual world. There is no way to ensure this except by stating a principle, and that's what she does. Her principle of **Event identification across worlds** states that, if an event occurs in worlds w_1 and w_2, and it is identifiable as having property P in w_1, it is identifiable as having P in w_2 as well.[40] Thus, in the case of (257), the speaker is saying that Jane has run in the actual world.

To complete the story, Hacquard must explain cases in which event identification doesn't apply. For example, why in (258) can we describe a event as a funeral in one world and a wedding in another? Event identification would seem to imply that, since it's described as a funeral in alternative worlds (those compatible with Bill's thoughts), the speaker is saying that it's a funeral in the actual world as well. Though I don't find her discussion completely clear, Hacquard's idea here seems to be that in this sentence, the event is described as a funeral not by the speaker, but by Bill. However, I don't understand in what sense Bill, but not the speaker, describes the event as a funeral. Bill thought that

---

[40] The final principle is a bit more complex than this because of the need to allow for actuality entailments with the Italian verb *volere* "want." However, these cases need not concern us here.

the actual event was a funeral, and of course the speaker doesn't think this. But the speaker says that it's a funeral in all of the accessible worlds (those representing Bill's thoughts). This is the same as the modal case. In (257), the speaker says that e_3 is an event of Jane running in some accessible world (a world compatible with Jane's abilities).

Let's set aside this problem and see how Hacquard explains the absence of the actuality entailment in imperfective sentences. She has a very interesting theory of imperfective morphology which claims that it can signal the presence of one of several operators. Among these operators are *Gen*, a progressive/habitual operator *Prog*, and a counterfactual element *CF*. Let's limit the discussion to *Gen* for simplicity. She builds on a quantificational theory of generics as in Heim (1982) and Krifka et al. (1995), among others. Here is her example and semi-formal representation of its meaning (her examples (127)–(128)):

(259)   (a)   A lion has a bushy tail.

       (b) $\text{Gen}_{[x,w]}(x$ is a normal lion in w, x has a bushy tail in $w)$

Notice that the generic operator binds both an individual and a world variable. It says that we should look across possible worlds for all of the normal lions, and that all (or virtually all) of these normal lions will have bushy tails. It is not just a statement about the normal lions in the actual world, since all of the actual lions could happen to have a property which is not characteristic of the species in a deep "generic" way.

I stress the point that the generic operator should quantify over possible worlds because the meaning Hacquard eventually adopts for the *Gen* operator does not. Simplifying somewhat, she proposes the following:

(260)   (a) $[\![\textit{Gen}]\!] (\alpha)(\beta) = \{w : \forall x[(\alpha(x) \land x$ is normal
from the perspective of $w) \rightarrow \beta(x)]\}$

       (b) $\{w : \forall x[(x$ is a lion in $w \land x$ is normal from the
perspective of $w) \rightarrow x$ has a bushy tail in $w]\}$

Notice that the only world involved in this formula is w, the evaluation (i.e., actual) world. This means that (259) will get the meaning in (260b) stating that all normal actual lions have bushy tails.[41] Hacquard needs a revised analysis of the generic operator.

---

[41] Hacquard (p.c.) agrees that *Gen* should quantify over non-actual individuals, but given the way her formalism works, (260) does not represent this meaning. α and β need a world argument other than w.

Suppose we were to fix Hacquard's semantics for the generic operator by adding quantification over possible worlds, as in the following:

(261)   (a) $[\![\, Gen \,]\!] \, (\alpha)(\beta) = \{w : \forall x \forall v[(\alpha(x, v) \wedge x$ and v are normal from the perspective of $w) \rightarrow \beta(x, v)]\}$

(b) $\{w : \forall x \forall v[(x$ is a lion in $v \, \wedge \, x$ and v are normal from the perspective of $w) \rightarrow x$ has a bushy tail in $v]\}$

While this leads to an appropriate meaning for the generic sentence, i.e. (261b), it fails to give the right meaning when combined with Hacquard's analysis of the ability modal. Suppose we give *Gen* scope over *Jane could-impv run*.[42]

(262)   (a)   *Gen*(Jane could-impv run)

(b) $\{w : \forall x \forall v[(C(x, v) \wedge x$ and v are normal from the perspective of $w) \rightarrow x$ in $v \, \wedge \, \exists w'[R(v, w') \, \wedge \, x$ is an event of Jane running in $w']])\}$

According to Hacquard, the value of *Gen*'s first argument α is given by context, and I represent this contextually given value as $C(x, v)$. In order to understand what predictions are made by (262), it is necessary to know what is entailed by this condition. In particular, does $C(x, v)$ entail that x exists in v? I think that it must. Hacquard says that α expresses the preconditions for event x. For example, preconditions for her running are that she's not asleep or sick, that there's no typhoon in progress outside, and so forth. Presumably what she means by saying that the preconditions for x are met is that they are met in v. (We certainly don't want to say that they are met in w, since that would give rise to an actuality entailment.) Given that x exists in v, and that x is an event of Jane running in w', the event identification principle kicks

---

[42] The meaning in (262) assumes that imperfective morphology does not contribute any temporal meaning. If we suppose that it indicates the opposite relation between t and x, we get the following:

(i)   (a)   *Gen*(Jane could-impv run)

(b) $\{w : \forall x \forall v[(C(x, v) \wedge x$ and v are normal from the perspective of $w) \rightarrow x$ in $v \, \wedge \, t \subseteq x \wedge t <$ now $\wedge \exists w'[R(v, w') \, \wedge \, x$ is an event of Jane running in $w']])\}$

This meaning runs into the same problems as (262).

An alternative to Hacquard's analysis in (262) would be to say that imperfective head doesn't undergo raising in the way that perfective does in (257), so that there is no actuality entailment to begin with. This would remove the motivation for the generic operator and therefore solve the problem we see in (262). It would also undermine her analysis of imperfective morphology, but, even so, this may be the best approach to preserving her central idea.

in. It tells us that x is an event of Jane running in v. Thus, according to Hacquard's theory (262) means that in every normal world (in which the preconditions for her running are met), Jane has the ability to run and (via the actuality entailment) in fact runs. The actuality entailment is still there. It has not been erased; rather, it has just been embedded under *Gen*.

With this explication of (262), it becomes apparent that this meaning is not correct. It says that in all normal worlds, Jane has the ability to run, and realizes this ability. This is not the same as having the ability to run. Because the actual world w need not be one of the the normal worlds v, we have no guarantee that she actually has the ability to run. All we know is that she has this ability in the normal worlds. And it won't help to assume (contrary to Hacquard) that the actual world is one of the normal worlds. The formula, in combination with event identification, implies that she runs in every normal world; so, if the actual world is a normal world, we get the entailment that she actually runs. But this is the the actuality entailment, and the point of introducing the *Gen* operator was to explain why imperfective sentences don't have this entailment.

The central problem with Hacquard's theory is that it is difficult to come up with a plausible meaning for *Gen* such that *Gen* + (*modality* + *actuality*) = *modality*. The above argument shows that *Gen* cannot be the operator we see in (259). Perhaps there is another kind of generic operator which solves this equation, perhaps the same kind that Bhatt had in mind with example (256). Nevertheless, simply finding one operator which cancels the actuality entailment is not sufficient for Hacquard's theory. She must show that any operator which licenses imperfective morphology leads to the correct overall meaning. Since one would expect the operator defined in (261) to license imperfective morphology, it seems that either her analysis of the actuality entailment or her theory of imperfective morphology will have to be changed.

Performativity and the actuality entailment

The actuality entailment remains an interesting open problem in the semantics of modality. To conclude our discussion of this problem, and of volitional modals generally, I would like to make two points. The first concerns the two-layer analysis of ability modals (various versions of which were given by Brown 1988, Hackl 1998, Bhatt 1999, and Horty 2001). It may well be that this analysis will play an important role in explaining the actuality entailment. However, if it is to do so, it must be extended beyond ability modals to the full range of cases which generate the actuality entailment, including various necessity and possibility priority modals. If we begin with Horty's account, it is easy to see in principle how to do so: he analyzes ability as $\Diamond$ plus *stit*, so we'd just allow various kinds of possibility and necessity operators in the place of the $\Diamond$. In practice, of course, the details are unlikely to be so simple. In any case, Bhatt's approach to using the two-layer semantics to account for the actuality entailment doesn't work, as Hacquard argues, essentially because he says that the overt morpheme which seems to represent the modal actually represents the *stit* (his *ABLE*). We'll need a new idea about how to use the two layers, and their interactions with aspect, to explain the actuality entailment.

The second point I'd like to make is that the actuality entailment can be classified as a kind of performativity. We defined performativity as a situation where a modal, by virtue of its conventional meaning, causes a root sentence it is in to perform an extra speech act in addition to that which is associated with its ordinary truth-conditional semantics. The actuality entailment is a case in point; the extra speech act is one of assertion.[43] The performativity of an ability modal can be represented in a dynamic framework as follows (choosing French *pouvoir* as an example):

(263)   **Update potential of *pouvoir*:** For any sentence ϕ of the form *pouvoir* ψ, the update potential of ϕ used in context c with modal base f and ordering source g, $[[\,\phi\,]]^{c,f,g}$, is defined as follows:

$$cg[[\,\phi\,]]^{c,f,g} = cg \cup \{[[\,\psi\,]]^{c,f,g}\} \cup \{[[\,\phi\,]]^{c,f,g}\}$$

A treatment of actuality entailment as an example of performativity can explain Bhatt's intuition that sentences with the actuality entailment contain an element similar to *manage*. According to this view, B's utterance in (252) asserts two things: first, that Yusuf has the ability to fly airplanes, and, second, that he did fly them. Why would B not

---

[43] We have seen two analyses of the actuality entailment. According to Bhatt's, it is not performativity, since for him sentences which show the actuality entailment do not contain a true modal. In contrast, according to Hacquard's theory, sentences which show the actuality entailment contain real modals, yet the entailment follows not from the compositional meaning (e.g., (257)) alone, but rather from this meaning in combination with the principle of Event identification across worlds. As I understand it, Event identification applies to the entire compositionally derived meaning and triggers an inference of actuality; if this is the right way to understand it, the actuality entailment can be seen as meeting the definition of performativity.

simply say that flew airplanes, since this entails that he has the ability to fly them? There must be a reason that B also asserted that he has the ability. One natural reason is that B fears A might doubt the simple statement because A is not convinced that he has the ability to fly. By saying both that he can and that he did, B both provides an answer and attempts to dispel A's grounds for doubting the answer. To put it another way, it only makes sense to go through this extra trouble of saying that he has the ability to fly airplanes if B thinks that A thinks he probably can't. Now, why would A think that Yusuf probably can't fly airplanes? Perhaps because flying airplanes isn't easy and Yusuf hasn't much training, or perhaps because there was a heavy storm. In such contexts as these, one infers that if Yusuf flew airplanes, it was despite such obstacles. This inference is close to the presupposition of *manage*.

I think it is progress to identify the actuality entailment as an example of performativity. Nevertheless, it is different in an important way from the performativity of epistemic and imperative-like deontic modals. In the latter instances, performativity appears to be a root phenomenon. In contrast, the actuality entailment persists in embedded clauses, as can be observed from the English example (264):

(264)   If John was able to finish the report, he'll be at the party.

Here the supposition of the *if* clause is that he did finish the report, not merely that he had the ability to. For this reason, I would hesitate to say that the actuality entailment is the same phenomenon as the performativity of epistemic and deontic modals in any deep theoretical sense. Nevertheless, there is at least a general similarity among the various types of performativity which is worth bringing out. It seems that, in certain circumstances, modals of all classes—epistemic, priority, and dynamic—introduce meaning into the sentence which goes beyond what can be explained in terms of their core modal interpretation. Future research in this area is likely to focus on the nature of the performative components of meaning and the issue of what causes these meanings to appear and disappear in various contexts.

4.4.2 Quantificational modality

We can define a quantificational modal as one which incorporates the semantics of an adverb of quantification together with some sort of additional, more properly "modal," meaning. By this criterion, the generic operator described informally in (259) (and perhaps formalized as in (261)) would qualify as a quantificational modal. A quantificational modal can be detected because, like an adverb of quantification, it gives rise to QUANTIFICATIONAL VARIABILITY EFFECTS. Notice the parallelism between (265) and (266):

(265) (a) A dog sometimes bites.
       (b) A dog always bites.

(266) (a) A dog can bite.
       (b) A dog will bite.

In (265a), the indefinite *a dog* seems to mean "some dog," while in (265b) it has a meaning closer to "all dogs." The fact that the quantificational force of the indefinite varies with the choice of adverb is the quantificational variability effect (QVE). There is a similar difference between (266a–b).[44]

As mentioned above, Lewis analyzed the QVE by hypothesizing that the indefinite introduces a free variable which is bound by the adverb, and his analysis is followed by later authors including Kamp (1981) and Heim (1982). An important alternative approach states that the adverb of quantification quantifies not over individual variables introduced by the indefinite, but rather over situations.[45] In this book, I do not have the space to review the differences between the two theories; let me just say that I believe that the situation-based theory was proven superior. If we want to analyze quantificational modals similarly to adverbs of quantification, we need to know how the situation-based theory works. Therefore, in what follows I will present a version of the situation-based analysis of adverbs of quantification.

---

[44] These sentences of course have other readings which should be ignored here. In general, *can* may have either a volitional or a quantificational reading, *will* may have either a quantificational or a future reading, and the indefinite subject may be interpreted either existentially or universally. The interpretations we are focusing on here are the one where the existential reading of the subject arises because it is associated with *can* and the one where the universal reading arises because it is associated with *will*.

In most cases, the irrelevance of the other meanings should be obvious, but I will comment on a couple. Example (265a) has a meaning similar to "All dogs sometimes bite." It is usually assumed that this reading results from a phonologically null operator giving *a dog* a universal reading, with *sometimes* having narrow scope and quantifying over times or situations. Example (266a) has a similar reading, "All dogs have the ability to bite," which can be given a similar analysis. In this case, a null wide scope operator gives the subject a universal meaning and *can* is a volitional (ability) modal, not a quantificational modal.

[45] For discussion, see Berman (1987); Groenendijk and Stokhof (1990); Heim (1990); de Swart (1991); Chierchia (1995b); von Fintel (1995), among many others.

To begin with, we need to know what a situation is and how it fits into semantics. There are several theories of situations and related entities (like events) which we might recruit as the basis of a theory of quantification, for example, the frameworks of Davidson (1980), Barwise and Perry (1981), or Kratzer (1989). There already exist well-developed analyses of adverbs of quantification based on Kratzer's version of situation semantics, and so in what follows I will follow her framework.

In Kratzer's theory, situations are basic entities, and situations may have other situations and individuals as parts. That is, we take as basic a set S of situations, a set A of individuals, and a part-of relation $\leq$ on $S \cup A$. When $x \leq s$, it is sometimes natural say that x is a part of s, while other times it is natural to say that x is in s. Possible worlds are maximal situations, that is situations which are not part of any other situation: w is a world iff $\neg \exists s [s \in S \wedge w \leq s]$. No situation is a part of more than one world. One can develop an intuition for Kratzer's notion of a situation by starting with the set of possible worlds, and then thinking of any part of such a world as a situation. These parts may be defined in spatio-temporal terms; for example, what's going on within one meter of your body as you read this sentence is a situation, as is everything which took place in Rome in 1949. Kratzer believes that there are other, more abstract situations as well, for example, one comprising just the redness of my copy of Palmer's *Mood and Modality*. Such abstract situations will not be important to us here; see Kratzer (1989) and Portner (1992) for discussion.[46]

In possible worlds semantics, a proposition is a set of possible worlds. In situation semantics, it is a set of situations. We say that a proposition p is true in s iff $s \in p$. In general, if we make the move from possible worlds semantics to situation semantics, we want to preserve the results of possible worlds semantics by making sure that, for any sentence ϕ, the set of worlds in which ϕ is true on the old theory is the same as the set of worlds in which it's true on the new theory. But because not every situation is a world, we also have to figure out in which non-world situations ϕ is true.[47] Here is Kratzer's meaning for a simple sentence from Kratzer (2007) (cf. her (34), simplified):

[46] Kratzer also assumes that individuals exist in only one world, but I will ignore the technicalities of this decision in what follows.

[47] Kratzer claims that there are two kinds of propositions within possible worlds semantics, generic and non-generic ones. (The sense of "generic" here may or may not be the same as the one other scholars wish to capture with the *Gen* operator.) A generic proposition is true in every situation of a world in which it's true, while a non-generic proposition is true in only some of the situations in a world in which it's true. We can ignore the generic propositions for the time being.

(267)   [[*There are two teapots*]] $= \{s : \exists s'[s' \leq s \wedge$ there are exactly two teapots x such that $x \leq s']\}$

Let us list some of the kinds of situations which are in the proposition [[*There are two teapots*]] :

1. The MINIMAL SITUATIONS: the smallest situations about which one could say they contain two teapots.
2. The MAXIMAL SITUATIONS: the possible worlds in which there are at least two teapots.
3. The OVERLOADED SITUATIONS: the situations which contain two teapots and more besides. (The possible worlds which contain two teapots are examples of overloaded situations.)
4. The EXEMPLIFYING SITUATIONS: the situations which contain two teapots and nothing else besides. (The minimal situations are all exemplifying situations.)
5. The COUNTING SITUATIONS: the spatio-temporally connected exemplifying situations which contain two teapots and nothing else besides, and to which you could not add any spatio-temporally connected situation and still have an exemplifying situation. ("Counting situations" is my term; Kratzer describes these as the maximally self-connected exemplifying situations.)

We can see the differences among minimal, exemplifying, and counting situations by considering a sentence with an activity verb (cf. her (37)):

(268)   Josephine flew an airplane.

Suppose Josephine flew an airplane twice, each time for two hours. The minimal situations are the smallest ones about which one could say that they contain her flying the plane. Maybe these last one second, or one millisecond. The exemplifying situations are all of the situations which you get by combining those minimal situations: the first five minutes of the first flight, the last ten minutes of the second, and so forth. Some exemplifying situation will be strange, because they combine the parts of the two flights, for example the first five minutes of the first flight and the last five minutes of the second. (See Kratzer 2007 for more on the complexities of exemplifying situations.) There are only two counting situations: the two complete two-hour flights.

The counting situations are important because they are often the natural units for semantics. For example, it is because there are two counting situations that *Josephine flew an airplane twice* is true. In particular, we can base our analysis of adverbs of quantification on them. I'll assume a function COUNTING which selects the counting situations from a proposition p. The following is a situation-based semantics for *always* based on Berman (1987), von Fintel (1984) and Kratzer (2007):

(269)   **Semantics for adverbs of quantification:**

$[\![\ always\ ]\!]\ (\alpha,\beta) = \{s : \forall s'[(s' \leq s \ \wedge\ s' \in$ $COUNTING(\alpha)) \rightarrow \exists s''[s' \leq s'' \wedge s'' \in \beta]]\}$

(270)   (a)   When Josephine flew an airplane, she always landed it safely.

(b) $\{s : \forall s'[(s' \leq s\ \wedge s'$ is a counting situation of Josephine flying an airplane) $\rightarrow \exists s''[s' \leq s'' \wedge\ s''$ is a situation in which she lands the unique airplane she was flying in s'' safely]]\}$

The meaning in (270b) can be paraphrased as "Each complete flight Josephine took can be extended to a situation in which she landed safely."[48]

With this review of the situation-based semantics for adverbs of quantification behind us, we can turn to quantificational modals. Both Heim (1982) and Brennan (1993) propose that they function simultaneously as adverbs of quantification and modals. Within situation semantics, saying they are like adverbs of quantification means that they quantify over counting situations. But, unlike adverbs of quantification, they are not limited to actual situations; they quantify over situations in other possible worlds besides. For simplicity, I will use a single accessibility relation R here, avoiding the complexities of Kratzer's ordering semantics, and assume that R is a relation between situations and worlds, so that $R(s, w)$ means that w is accessible from s. The notation w_s indicates the world of which s is a part. Given these conventions, we get the following meanings for *will* and *can*:

---

[48] By the way, (270b) demonstrates how pronouns are treated within the situation-based theory; they are definite descriptions which pick out a unique referent relative to the situation.

One further issue may arise from the need to ensure that counting situations from α are matched with the situations in β in the right way. We wouldn't want to say that (270a) is true if she only landed safely the first time, even though one could extend the counting situation which represents the second flight to one which includes the first flight, and this would be one in which she landed the plane safely. There are various ways to solve this problem, and I'll set it aside for now.

(271)   **Situation semantics for quantificational *can*:**
$$[\![\text{ can }]\!]\,(\alpha,\beta) = \{s : \exists s'[R(s, w_{s'}) \wedge s' \in \text{COUNTING}(\alpha)$$
$$\wedge\ \exists s''[s' \leq s'' \wedge s'' \in \beta]]\}$$

(272)   **Situation semantics for quantificational *will*:**
$$[\![\text{ will }]\!]\,(\alpha,\beta) = \{s : \forall s'[(R(s, w_{s'}) \wedge s'$$
$$\in \text{COUNTING}(\alpha)) \rightarrow \exists s''[s' \leq s'' \wedge s'' \in \beta]]\}$$

Let's look at two concrete examples. First, one with *can*:

(273)   (a)   If a dog is angry, it can bite.
        (b) $\{s : \exists s'[R(s, w_{s'}) \wedge s'$ is a counting situation of a dog being angry $\wedge \exists s''[s' \leq s'' \wedge s''$ is a situation in which the unique dog in s'' bites]]\}$

Suppose that we have a circumstantial accessibility relation such that $R(s, w)$ holds iff w is a world in which dogs have the same physical and psychological make-ups that they have in s. In that case, the sentence can be paraphrased as "Some possible situation of a dog being angry (in a world where dogs are pretty much like they really are) can be extended to a situation in which that dog bites you." In other words, "Some angry dog bites you in a world where dogs are pretty much like they are in this one."

Now let's see a similar example with *will*:

(274)   (a)   If a dog is angry, it will bite.
        (b) $\{s : \forall s'[R(s, w_{s'}) \wedge s'$ is a counting situation of a dog being angry $\rightarrow \exists s''[s' \leq s'' \wedge s''$ is a situation in which the unique dog in s'' bites]]\}$

Assuming the same accessibility relation R, this meaning can be paraphrased "All possible situations of a dog being angry (in a world where dogs are pretty much like they really are) can be extended to a situation in which that dog bites you," or "All angry dogs who live in worlds where dogs are pretty much like they really are bite you." This analysis seems close to correct, but it is too general. The sentence does not really entail that all those angry dogs bite you—only that all of those angry dogs who have an opportunity to bite you do so. This shows that the universal quantification over situations in (272) must be contextually restricted. Since contextual restriction is a property of all natural language quantifiers, we can set aside the issue of how to implement such a restriction here.

When it comes to simple examples lacking an overt *if* clause, like those in (266), we can either assume that context fills in the relevant material or that the first argument of the modal is provided in some other way. For example, it could be that the indefinite itself serves as the argument:

(275)   can(a dog$_i$, t_i bites)

For this latter analysis to work in the context of situation semantics, the indefinite would have to express a proposition. For example, *a dog* on its own would mean "There is a dog." Moreover, the trace t_i would have to be interpreted as a pronoun. Putting these assumptions together, the sentence would mean "Some dog in a possible world where dogs are pretty much like they really are bites somebody," which seems correct.

A final issue in the situation semantics analysis of quantificational modals is what kinds of accessibility relations are available. We find volitional-like meanings with quantificational modals, but we do not find priority-like or epistemic-like meanings:

(276)   (a)   (i)   A wild dog can/will bite.
              (ii) A tall person can/will see the ocean from here.
        (b)   A wild dog can/will be sent to the pound.
        (c)   #A wild dog can/will be sleeping outside.

Example (276a) is quantificational (note the QVE when we switch from *can* to *will*) and the meanings are very similar to those we would get with volitional modals. In (276b) I try to give similar examples with deontic meanings. At first glance it may seem that these examples mean "Some/all situations containing a wild dog and in which all relevant regulations are adhered to can be extended to situations in which the dog is sent to the pound," but I don't think this is quite right. A better description of the facts is that *can* has a deontic reading and *will* a future (or perhaps dispositional/willingness) reading. When we turn to (276c), it is clear that *can* has no epistemic reading.

Based on these considerations, I suggest that quantificational modals are lexically restricted to the same kinds of accessibility relations (or conversational backgrounds) as volitional modals. This might explain why quantificational and volitional modals are expressed by the same forms, *can* and *will*. Still, we cannot unify volitional and quantificational modals. One difference is that the former quantify over possible worlds, while the latter quantify over situations, and although this contrast might be seen as rather minor, there is another difference as well.

Recall that volitional modals take their subject as a semantic argument in the fashion of control predicates, so that an ability modal, for example, will have to do with the abilities of the subject. But, as we have seen in (244d), quantificational modals can occur in sentences with expletive subjects, and so they are not in general control predicates. Future work on quantificational modals is likely to focus on the relationship between their modal meaning, closely connected to volitional modality, and their adverbial meaning, based on a quantification over situations and designed to explain the QVE.

4.5 Looking ahead

In this chapter, we have examined the various categories of sentential modality with the aim of understanding the most important issues which are currently affecting the evolution of semantic theories of modality. There is a great deal of exciting research going on in this area, at both the empirical and the theoretical levels, and I hope that the chapter has served to to establish some conceptual order in what can easily seem an overwhelming variety and amount of scholarship.

Thus far, we have looked at modal expressions with the goal of understanding the modal meanings they express. This may seem the obvious thing to do, but in fact it has led us to ignore a great deal. In particular, we have not focused on the complex ways in which modals interact with other linguistic categories or the fact that non-modal components of meaning are often incorporated into their interpretations. In the next chapter, we will look at some of the most important ways in which modality becomes intertwined with other semantic categories.

5

Modality and Other Intensional Categories

In previous chapters, we've taken the strategy of trying to isolate modality from other semantic categories. The purpose of this strategy is to look carefully at the nature of modality in order that we can develop a precise, explanatory theory of it. However, in reality modal semantics is intertwined with many other topics, and in the following sections we will investigate several:

5.1 Temporal semantics, including tense and aspect
5.2 Conditionals
5.3 Mood and evidentiality

The first topic we'll study in some depth. The second we will examine in a more cursory way, because the literature on conditionals is so vast that another entire volume would be needed. The third will be treated in an even more superficial way, because an entire volume on it is due to appear (Portner, forthcoming).

5.1 Modality and time

I'd like to divide up the ways in which there is an important relationship between modality and time into two main areas. First, as we try to understand elements like modal auxiliaries, whose meanings are primarily in the modal domain, we find that they affect the temporal interpretation of the sentences they occur in as well. I discuss the temporal semantics of modal elements in Section 5.1.1. And, second, as we try to understand the semantics of tense and aspect morphemes, whose meanings are traditionally conceived as having to do with time and event structure, we see that they also introduce modal meaning into the sentence. In Sections 5.1.2–5.1.3, I will focus on modality in the semantics of tense and aspect.

5.1.1 The temporal orientation of modal sentences

Sentences containing modal auxiliaries sometimes have to do with past situations, sometimes with present situations, and sometimes future situations. This is the case for epistemic, priority, and dynamic modals.

(277)   Epistemic

  (a) He might have left.
  (b) He might be in the other room.
  (c) He might leave soon.

(278)   Priority (here deontic)

  (a) He should have left.
  (b) He should be in the other room.
  (c) He should leave soon.

(279)   Dynamic (here ability)

  (a) (i) He could swim.
    (ii) He could have swum.
  (b) He can swim.
  (c) (i) #He can swim tomorrow.[1]
    (ii) He will be able to swim tomorrow.
    (iii) He will swim tomorrow. (volitional[2])

There has been some very interesting research into various aspects of the problem of temporal interpretation of modals, but—even limiting our attention to English—no one has presented a theory which aims to cover all of the cases.

---

[1] There is a tendency for volitional sentential modals not to allow future readings, as with English *can*, but such a meaning is not always ruled out. For example, in Mandarin *hui* "can, will" only means "will" in simple future sentences, but in the following it has a future ability meaning:

(i) Xiaoming, xianzai ni hai bu hui shang cesuo, danshi mingnian ni jiu
  Xiaoming now you still not can use the toilet but next year, you then
  hui le!
  can PART
  'Xiaoming, you still can't use the potty, but next year you'll be able to.'

We lack a good explanation of the fact that volitional modals resist future interpretation.

[2] This is a relevant example if we consider *will* to express voltional modality.

An analysis of the temporal interpretation of modal sentences must determine the contributions of three types of factors: first, any independent tense or aspect operators which may be in the sentence, and their scope properties; second, the temporal interpretations of the modals themselves; and, third, general semantic or pragmatic principles which help determine temporal meaning, but which are not tied to any particular grammatical element. I will first examine the role of tense/aspect operators, and then turn to the meanings of the modals themselves. I'll mention the possible contribution of general principles as it becomes relevant during the discussions of temporal and modal operators.

Past interpretations: temporal operators in combination with modals

There are two kinds of temporal operators we need to consider, tense and aspect. As far aspect goes, modals can combine with both the progressive and the perfect.

(280)   (a)   She might be winning.
        (b)   She might have won.

No special issues seem to arise with the progressive, but the role of the perfect in modal sentences is controversial and complex, as we'll see below.

In English, it is difficult to see whether tense occurs in a modal sentence. (For this reason, it's unfortunate that English is the language on which most relevant theoretical research has been done, but see Mondadori 1978, and Hacquard 2006.) It is traditional to divide modals into pairs which differ only in terms of whether a past tense morpheme is present, as follows:

(281)   (a)   [PAST] could, should, might, would
        (b)   [PRES] can, shall, may, will, must

The idea that the modal in some cases contains a tense morpheme is supported by the tense morphology in embedded clauses. In many languages, a clause embedded in a past tense matrix will itself have past tense morphology, but this morphology is not interpreted. This phenomenon is known as SEQUENCE OF TENSE.

(282)   (a)   Mary said that John was her husband.
            ≈ "Mary said 'John is my husband (now)'.", or
            ≈ "Mary said 'John was my husband (before)'."

(b)   Mary said that John could swim.
    ≈ "Mary said 'John can swim (now)'. ", *or*
    ≈ "Mary said 'John could swim (in the past)'. "

Example (282a) has two interpretations, indicated by the two para-phrases. On the first interpretation, called the "simultaneous reading," Mary said at some past time t that John was then, at t, her husband. This interpretation is often analyzed in terms of the idea that the past tense on the embedded clause, here *was*, does not contribute to the seman-tics; it re-expresses the past tense of the main clause in the embedded clause, and the embedded clause is itself tenseless. In contrast, on the second "shifted" reading the embedded clause tense is thought to be semantically real: she said at some past time t that John had been her husband *before t*. We see in (282a) a complex pattern, and several proposals are available in the literature for how to explain it (e.g., Enç 1987; Abusch 1988, 1997; Ogihara 1989, 1996). The key point here is that, in such contexts, ability *could* behaves like the past tense of *can*, as seen in (282b). This fact strongly suggests that volitional *could* involves a past tense.

The other modals in (281a) present a more confusing picture. *Should* is representative:

(283)   Mary said that John should leave.

While (283) has the simultaneous reading, according to which John's obligation holds at the time of Mary's saying, it lacks the shifted read-ing. This point suggests that *should* contains a past tense which can re-express the past tense of the main clause, but it cannot function as a semantically real tense.

The difference between *could* and *should* also surfaces in root clauses. *Should* can be used to describe a present obligation but not a past one. On the other hand, ability *could* must concern the past.

(284)   (a)   She should leave. (deontic reading)
    (b)   She could swim. (ability reading)

Thus it seems that modal sentences sometimes involve tense. Some of the time this tense is grammatically present but not semantically active, while in other instances it is both grammatically and semantically real. As we get further into the complexities of temporal interpretation, we'll have to identify those semantically active tenses and distinguish them from the tenses which are realized in morphology but semantically inactive.

Next we turn to the issue of scope. As was mentioned in Section 4.2.1, it is often said that epistemic modals cannot take scope under a temporal operator. (285) contains an overt operator, the perfect form, which contributes some sort of past meaning to the sentence; nevertheless, it concerns the speaker's knowledge at speech time, not at a past time such as that when Mary is thought to have left:

(285)   Mary must have left.

(285) is interpreted with the scope "must(PERF(Mary leaves))." On the basis of examples like this one, many scholars have stated that the temporal interpretation of epistemic modals is directly determined by context, without the intervention of any tense or tense-like operators (e.g., Groenendijk and Stokhof 1975; Abusch 1997).

The assumption that epistemic modals never have scope under a temporal operator has been challenged. As noted in Section 4.2.2, von Fintel and Gillies (2007b) argue that the modal in (207), repeated here as (286a), has scope under tense. A similar example, (286b), is given by Condoravdi (2002, (6b)).

(286)   (a)   The keys might have been in the drawer.
         (b)   He might have won the game.

This latter example has a reading in which we're concerned with a past possibility; it can be brought out with the continuation *but he didn't in the end*. Condovardi says that this reading does not exemplify epistemic modality, but rather metaphysical (or historical) modality. That is, on the reading in question (286b) is supposed not to have to do with what the speaker knew at that point, but with what futures were still open at that point. However, (286a) clearly has an epistemic meaning. Note that examples like (286a) are very difficult to construct with a non-stative clause.

Non-epistemic modals can easily be interpreted with respect to a past time:

(287)   (a)   Jane should have taken the train.
         (b)   In those days, I could run.
         (c)   *In those days, I can (have) run.

We should note some patterns within the English data here. A priority modal like the teleological or deontic (287a) can be concerned with a past time only when it has perfect morphology (*have taken*). In this respect, it is similar to the epistemic/metaphysical examples (286). In contrast, some dynamic modals, like that in (287b), can have a past interpretation even without the perfect; we might attribute this difference to the idea that *could* contains a semantically active past tense. Other modals cannot have a past interpretation at all, as in (287c)

A further complexity is that certain combinations of a modal with tense or aspect forms result in counterfactual interpretations. In English, we typically find counterfactual meanings when a modal with past tense morphology is combined with the perfect, as in (287a); this example implies that Jane did not take the train. Likewise, on the relevant reading (future possibility in the past), (286b) implies that he did not win, as pointed out by Condoravdi. To explain this counterfactuality, Condoravdi (2002) proposes that counterfactuality arises as a conversational implicature based on the idea that the perfect aspect has scope over the modal. That is, the representation of (286b) is "PERF(MIGHT(he won the game))." Mondadori (1978) and Hacquard (2006) make similar proposals in which a past operator takes scope over the modal, but the details are different. For example, Hacquard, who concentrates on Italian examples, proposes that these examples contain a covert counterfactual modal, decomposed into a combination of the past tense and a future modal, in addition to the overt modal.

We will focus on Condoravdi's analyses here, since it has been developed in the most detail. Recall that she thinks that the modal in (286b) has a metaphysical or historical meaning. She then notes that, at any time t, the range of futures still open at t is a subset of those open at any time t' which precedes t. Therefore, the perfect form is semantically weaker than the non-perfect form. Why would a speaker choose to use the weaker perfect form instead of the stronger non-perfect form? Condoravdi's answer is: because the stronger form is not true. That is to say, it is no longer possible for him to win the game.

I'll point out three issues. First, if counterfactuality arises as an implicature, we expect that it can be cancelled, but it is not so clear that it can be (?*At that point, he might have won the game, and in fact he did.*) Second, since the explanation is based on the modal's being interpreted with respect to a past time, if it can be extended to priority and dynamic modals, it would predict that the past use of *could* in (287b) should be counterfactual; however, counterfactuality only arises when the perfect form is present. This fact suggests that the perfect form itself is crucial to triggering the counterfactual meaning in English. And, third, the explanation depends on the modal having the force of possibility, since, with a necessity modal, the past form would be stronger; therefore, it would be difficult to extend this to Condoravdi's explanation to (287a).

Let us set aside the issue of counterfactuality and return to the past meaning. We know that there is some type of "pastness" in a variety of examples, but what does this pastness amount to? There are really two separate kinds of pastness that need to be kept in mind. One is the TEMPORAL PERSPECTIVE, to use the terminology of Condoravdi (2002). In possible worlds semantics, a temporal parameter is used by the modal to determine the set of accessible worlds, and this parameter usually comes from the context c. For example, in (285) we employ the accessibility relation R which holds between w and v just in case everything the speaker of c knows in w at the time of c holds in v as well. But in (286), we don't want to use the time of c, but rather some time before c. This type of shift is rare in the case of an epistemic modal (and it may be limited to stative sentences, as pointed out above), but it does occur.

We also find a past temporal perspective in non-epistemic cases, for example (287a–b). Thus, (287a) can mean "in light of Jane's goals then, she should have taken the train." It is less clear whether those examples which allow past perspective also allow present perspective. The following contexts bring out the relevant considerations:

(288)   At the time Jane wanted to enjoy some time alone in the car, so she decided to drive to her friends' house. But once she got there, she had so much fun that she wished she had taken the train so she could have arrived sooner. So, *she should have taken the train.*

(289)   Jane drove her own car to her friends' house, but on her way the road was washed out by a storm, and she was late. So, *she should have taken the train.*

In (288) one could say that *she should have taken the train* is true because we take into account her present, rather than her past, goals; that is, we could say that the modal takes a present perspective. In (289), one could say that it's true because we take into account a circumstance which holds now, but which was not yet true at the time she decided to drive; that is, the modal takes a present, not a past, perspective. If we look at the two examples in terms of Kratzer's theory, one interesting difference is that the first seems to involve present perspective of the ordering source (we look at her present goals), while the latter involves present perspective of the modal base (we look at present circumstances).

I'm not convinced that (288)–(289) illustrate present perspective. Consider (288) first: all along, Jane has the goal of enjoying herself as much as possible. At the time, she thought that driving was the right way of meeting this goal, since driving alone gives her a fair amount of pleasure. However, it turned out that spending time at her friends' place gave her even more pleasure. So, based on the the goals she had at the time, she should have taken the train. And next (289): from before the time Jane decided to drive, it was a fact that an accident would occur (though nobody could know this fact at the time). Thus, based on the facts which held at the time, she should have taken the train. What seems to be a non-past perspective here is actually a past perspective using a modal base which takes into account an unknowable, then-future fact.[3] If this way of looking at (288)–(289) is correct, the apparent differences in temporal interpretation have only to do with differences in the conversational backgrounds which serve as modal base and ordering source.

Besides temporal perspective, we have another kind of pastness, called TEMPORAL ORIENTATION by Condoravdi. This concerns the temporal relation between the speech time and the time at which the core clause, under the scope of all temporal and modal operators, is true. In (285), the hypothetical leaving event is located prior to the speech time. All of the examples in (285)–(287) have past temporal orientation (setting aside the ungrammatical (287c)). What we observed in our initial discussion of those examples was that epistemic and priority modals require the perfect form in order to get a past temporal orientation, whereas certain dynamic modals are able to have past temporal orientation on their own.

As one develops a theory of past interpretations of modals, the challenge is to find the right way of shifting the temporal perspective and temporal orientation. Focusing on epistemic and metaphysical interpretations, Condoravdi (2002) proposes that past perspective comes about when the perfect takes scope over the modal. She claims that this analysis is possible for metaphysical modals but not for epistemic ones, and that when a metaphysical modal has past perspective, there is no requirement of past orientation, as illustrated by (209c), repeated here:

(290)   He might have been available yesterday/next month.

---

[3] Controversy about whether there are any true future facts can be traced as far back as Aristotle. See Kaufmann (2005a) and Werner (2006) for recent discussions which are directly applied to the issue of temporal semantics.

She analyzes past orientation as arising when the perfect has scope under the modal, as in (285).

Condoravdi's analysis of metaphysical modals appears to apply to priority modals too. Example (287a) can have past perspective, in which case the orientation is free, as seen in (291), or past orientation, which case the perspective is present

(291)   Jane should have taken the train tomorrow instead of today.

If von Fintel and Gillies (2007b) are right that the epistemic (286a) has past perspective, we would expect that its orientation can be present or future. As pointed out in Section 4.2.2, Condoravdi thinks that such examples always have past perspective. If she is right, a different analysis of the epistemic examples is needed; in particular, the sentence will need two sources of pastness, and this is problematical since only one is overt, the perfect. Perhaps on this reading *might* contains a semantically active past tense in the same manner as dynamic *could*, though this hypothesis can be called into doubt because *might* never has a past interpretation in the absence of the perfect. It seems that a more innovative analysis of the epistemic examples is called for.

Some sequence of tense facts as discussed by Abusch (1997) are relevant here. She shows that modal expressions have past perspective when they are under the scope of a past tense verb:

(292)   (a)   My wife might become rich. (Abusch 1997, (39))
        (b)   John said that his bride might become rich. (based on Abusch 1997, (42b))

Example (292a) shows that unembedded *might* is interpreted with respect to the utterance time, as expected; (292b) shows that embedding it under *said* shifts it into the past.

At this point, it is worth remembering some of the arguments for relativism (Section 4.2.4). Recall that the meaning of an epistemic modal can depend on the knowledge of an individual other than the speaker, in particular when the modal is embedded under certain verbs (e.g., *say*). This shows that the "judge" parameter, to use Stephenson's (2007) terminology, is not a simple indexical. But it's not clear exactly what it is. (I suggested that it may be something like a shiftable indexical.) Clearly, the question is not yet resolved, but the point here is that we face a similar issue in understanding the temporal interpretation of modals. Is the time parameter employed by modals controlled by a temporal operator (tense or aspect), by context, or by a another type of operator? Given the complexities of the data, we should not expect that the same answer will be correct for all cases. For example, it could be that the temporal parameter of an epistemic modal may be shifted by context while that of a dynamic modal can only be shifted by tense.[4]

It is obvious that we don't yet have a good understanding of what happens when a modal is combined with temporal operators. Epistemic, priority, and dynamic modals all behave differently, and these differences seem to have to do with morphology, scope, and perhaps pragmatic factors. In addition, as yet there is no good analysis of the counterfactuality seen in many examples.

Present and future interpretations of modal sentences

Next we turn to the distinction between present and future interpretations of modal sentences. A sentence with present perspective can have present or future orientation, as in the following:

(293)   (a)   John might be here.
        (b)   John might leave.

These are epistemic examples. Data illustrating future interpretations of sentences expressing priority and dynamic modality are given in (278)–(279).

Putting aside the case of *will*, theoretical discussions of the temporal interpretations of modal sentences seem to agree that tense is not involved in the future readings. Enç (1996) attributes futurity to the modals themselves; according to her, some of them incorporate a tense-like semantics into their meaning. In the following, note the shift of the time parameter from i to i' (from Enç 1996, (33)):

(294)   MODAL[S] is true at $\langle w, i \rangle$ iff in every world w' accessible to w there is an interval i' such that $i < i'$ and S is true at $\langle w', i' \rangle$.

Enç attributes this meaning to deontic *must*, and it appears that she intends the same general approach to apply to all cases in which a sentence with a modal auxiliary has a future orientation.

Other analyses of the future interpretations of modal sentences claim that futurity is due to an interaction between the modal meaning and

---

[4] These observations suggest another way to look at von Fintel and Gillies's (286a). Perhaps the temporal parameter here is shifted not under the influence of a temporal operator, but rather on the basis of context alone. If this is right, the construction could be thought of as similar to free indirect style (on which, see Sharvit 2004 for a recent theoretical discussion).

general principles. Both Condoravdi (2002) and Werner (2006) dispute Enç's proposal that futurity comes solely from the modal itself, though these two authors make proposals which are quite different in other respects. Condoravdi makes the point that future readings are typically obligatory with event sentences and optional with state sentences (examples modified from her (1)):

(295) (a) He must/ought to/should/may/might get sick. (event)
       (b) He must/ought to/should/may/might be sick. (state)

Based on this correlation with aspectual class, Condoravdi builds a theory which aims to explain the future readings of modal sentences.

Condoravdi's theory is based on two components: a definition of what it is for an event or state to be true at a time, and a specification of the temporal semantics of modals. As with much other work on temporal semantics, she assumes that untensed sentences denote properties of events or states. For example, in (295a), the constituent *he gets sick* denotes the property which is true of an event e at world w iff e is an event of him being sick in w. In terms of such properties, she gives the following definition (modified from her (19)):

(296) $AT(t, w, P) =$

       (a) $\exists e[P(w, e) \wedge \tau(w, e) \subseteq t]$, if P is eventive.
       (b) $\exists e[P(w, e) \wedge \tau(w, e) \circ t]$, if P is stative.

Here, $\tau(w, e)$ is the time interval occupied by e in w. Basically, this says that, for an event/state to be AT a time, the event must be included within the time, while the state merely needs to overlap it. This principle is not meant to be specific to modal sentences. A principle like this one is needed for variety of temporal phenomena; see Portner (2003) for discussion.

Condoravdi also makes a proposal concerning the lexical semantics of certain modal auxiliaries. Her discussion is limited to epistemic and metaphysical modals; within this class, she gives an analysis similar to that of Enç (1996) and Abusch (1998) on which the modal shifts the evaluation time of the clause under its scope. More precisely, she proposes that these modals shift the evaluation time to an interval which stretches from the reference time (the speech time in modals not under the scope of a temporal operator) to the end of time. This interval is indicated as $[t, \_)$. In combination with the definitions in (296), her proposal accounts for the role of aspectual class in (295). Because an event must be included within $[t, \_)$, it will have to be in the future; in contrast, a state need only overlap $[t, \_)$, and so the only restriction is that it not be completely in the past.

Werner links the choice between present and future readings to the distinction between epistemic and non-epistemic interpretations. His fundamental goal is to derive the future interpretations of modal sentences using the difference between an epistemic and a circumstantial modal base. Epistemic modals easily have a present interpretation, but non-epistemic modals do not, as illustrated in (297), his (1), (3):

(297)　(a)　Jim might be late. (epistemic)
　　　　(b)　Jill may be seated. (deontic)

However, the correlation has exceptions, as illustrated by (278). The point may be more clear in the following:

(298)　He should be in the other room, but he's not. He's still out with his friends.

Werner avoids dealing with such cases, limiting his attention to the following modals: *may*, *might* ("on its non-past, non-counterfactual readings," p. 238), *will*, *can*, *must*, and *shall*. This mostly avoid the exceptions, though dynamic *can* remains a problem. Because she limits her attention to epistemic and metaphysical modals, Condoravdi does not account for this pattern.[5]

In order to evaluate Werner's work, we have to consider whether the modals he considers should indeed be separated out as a special class. On the one hand, he points out that, except for *might*, they all lack past tense morphology. This suggests that, if we exclude *might*, they may indeed form a natural class for temporal semantics. But, on the other hand, there are reasonable alternative explanations for the pattern which he finds. Let's begin with the assumption that a modal will be compatible with both present and future meaning unless something prevents one of these possibilities; thus, there's nothing to explain with the epistemic examples. Ninan (2005) gives an account of the lack of

---

[5] Condoravdi's analysis of the scope relations between the perfect and an epistemic or metaphysical modal deals with a similar pattern: the temporal orientation of a metaphysical modal must be future, while that of an epistemic modal can be past, present, or future. Werner's explanation of the contrast in (297) is very similar to Condoravdi's account of the difference between epistemic and metaphysical modals.

In evaluating Condoravdi's analysis, Werner claims that Condoravdi derives the modal interpretation from the temporal, while he does the opposite, but this point is not based on a fair comparison; it is simply due to the fact that he does not attempt to account for the role of aspectual class in the temporal interpretation of modal sentences.

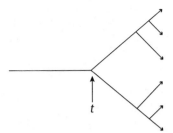

FIGURE 5.1. A set of histories which are alike up until time t

present and past interpretations for deontic *must* and *may* based on performativity, as discussed in Section 4.3.3. It is unclear whether non-epistemic *will* and *shall* are modals, as opposed to tense markers. If they are tense markers, they are not relevant here; if they are modals, their futurity may follow from the particular type of modality they express, as argued by Enç (1996). This covers all of the cases. Thus, perhaps Werner's pattern is due to the particular range of cases he chose to examine.

Werner's analysis is based on the BRANCHING TIME MODEL of the relationship between time and possible worlds. The intuition behind this model is that time has the structure of a tree; at any point, there is a single path into the past (towards the root) but multiple open futures. A complete history, a path upwards from a root, corresponds to a possible world in more ordinary possible worlds semantics. See Figure 5.1. Dowty (1977) and Thomason (2002) contain discussion and formal definitions. Branching time is one way of conceptualizing the historical modal base discussed in Section 2.3.3.

Given this branching time model, Werner proposes that all non-epistemic modals use a historical modal base f defined as follows:

(299)   For any world w and time t, $\bigcap f(w, t) =$ the set of worlds which are identical to w up through time t.

Werner relates this historical modal base to Kratzer's totally realistic modal base. One could say that this is a temporally restricted version of the totally realistic modal base, and in fact he calls the f in (299) "totally realistic."

The fact that non-epistemic modals have the modal base in (299) implies that they access only histories which branch from our own past. In contrast, an epistemic modal may access a wider range of histories. This difference is illustrated in Figure 5.2. Given these assumptions, Werner's analysis is based on two principles. One principle (his

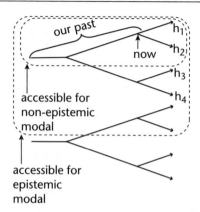

FIGURE 5.2. Werner's analysis

Disparity Principle) states that, given a modal sentence $M(S)$, S is true in some worlds compatible with the modal base and false in others. Therefore, if a necessity sentence is true, the ordering source must play a crucial role, making it the case that not all worlds compatible with the modal base are accessible. For a non-epistemic sentence, this condition will be met if S is true in h_1 and false in h_2, while for an epistemic sentence it can be true in h_1 and h_2, and false in h_3. The second principle (the Non-disparity Principle) states that the truth value of S cannot vary among worlds which branch from one another after now. Thus it implies that S is true in both h_1 and h_2, or that it's false in both. In the case of non-epistemics, these principles are in conflict. He assumes that the principles are ranked and violable, and that when there is a conflict the first one wins out.

Now consider the epistemic (297a). If this sentence is true and the principles are satisfied, either Jim is late in h_1 and h_2 and not late in h_3 and h_4, or he is late in h_3 and h_4 and not late in h_1 and h_2. Thus, his lateness is already settled as of now; in other words, we have a non-future interpretation of the modal. In contrast, with the deontic (297b), either he is seated in h_1 and not seated in h_2, or the other way around. This implies that he sits down, in any accessible world in which he does sit down, in the future.

Two important issues arise for Werner's approach. First, his account implies that epistemic modals do not have future interpretations, and, as a result, he claims (if I read him correctly) that examples which appear to contain an epistemic modal with a future interpretation are actually not epistemic. Rather, they involve a totally realistic modal base and incorporate epistemic-like factors into the ordering source. This account leaves unexplained why the same forms are often used for the present "true epistemics" and future "epistemic-like" meanings. But in favor of his proposal, one can note that sentences with epistemic *must* do not have a future interpretation (except for the "scheduling reading" of *He must leave tomorrow.*, on which more below). This cannot follow from aspectual class, since *must* combines with stative sentences, which should support both present and future readings, as it does with *may*:

(300)   (a)   John may be happy when he gets here tomorrow.
         (b)   #John must be happy when he gets here tomorrow.

In fact, there are no strong epistemic(-like) future modals in English. Werner asserts that his theory accounts for this fact (he phrases it as the prediction that future *will* lacks an epistemic interpretation), but I do not understand his argument. In any case, the lack of future interpretation for (300b) is not explained by Condoravdi's aspectual analysis.

A second problem for Werner's theory is due to the fact that, according to him, all non-epistemic modals use the same modal base, the temporally restricted totally realistic one, and that these modals differ only in the choice of orderings source (for which he suggests ones based on normative, physical, and biological laws). The problem here is that different readings of these modals appear to be based on different sets of circumstances. For example, the difference between intrinsic and extrinsic circumstances is often relevant for dynamic modals:

(301)   (a)   In view of the fact that it's made of porcelain, the vase can be broken.
         (b)   In view of the fact that it's in a vault, the vase cannot be broken.

One might suggest that these differ in the ordering source. Perhaps the ordering source (301b) contains the proposition that the vase will stay where it is, while that in (301a) does not. This solution works technically, but the assumption that the vase will not stay where it is doesn't seem to be needed to make (301a) true. Rather, it just seems that we limit our circumstantial modal base to intrinsic circumstances.

We have seen two approaches to the contrast between present and future temporal orientation of modals. The link Condoravdi draws between temporal interpretation and aspectual class seems clearly correct, and must be a part of our ultimate explanation. Werner aims to derive temporal interpretation from modal type, and, while the attempt is interesting, it has some difficulties; the lack of strong epistemic modals with future meaning provides the best argument for his general perspective, but the details are not clear. More generally, both Condoravdi's and Werner's theories suffer from the fact that they only take into account a limited range of data, and much remains to be done to bring the full pattern into focus.

5.1.2 The modality of tense

Studies of tense and aspect have shown that modal meaning is often involved in the semantics of forms whose meanings primarily have to do with time. In this section and the next, I will briefly survey some of the connections, focusing first on tense. Because the semantics of tense is itself a complex and difficult topic, it will not be possible to give a careful and detailed discussion of the relevant phenomena. Instead, I just hope to trigger an appreciation of the wide range of cases in which modality becomes relevant to the study of tense.

The future and *will*

The connection between tense and modality is most obvious with expressions that talk about the future. From a certain perspective, it would be surprising if there were not a link between these topics. The branching time model makes precise in an appealing way the intuition the future is not yet determined, in contrast to the past. According to this perspective, the future is automatically a kind of modality. In order to talk about the future, we must quantify over the future parts of histories; since histories are, in effect, worlds, any operator which lets us talk about the future will be very similar to modals as analyzed in possible worlds semantics.

We can also think about the connection between the future and modality from a grammatical perspective. What is the category of *will*? On the one hand, because it is grammatically like modal auxiliaries, it is natural to suggest that *will* expresses a kind of modality; if English is representative of other languages, one might further conclude that the future is in general a kind of modality. But, on the other hand, many languages have future forms which are not so similar to modals, but which are more similar to other tense forms; taking such languages as representative, one might conclude that the future is expressed by a tense in general, and that *will* is the form which English uses for this job. Large numbers of scholars line up behind both positions.

(See Montague 1973; Wekker 1976; Bennett and Partee 1978; Comrie 1985; Hornstein 1990; Kamp and Reyle 1993; Gennari 2003, for the first view; and Joos 1964; Leech 1971; Lyons 1977; Palmer 1986; 1990; Dowty 1979; Quirk et al. 1985; Enç 1996; Huddleston and Pullum 2002; and Kissine 2008, for the second). The recent theoretical literature takes the stand that *will* is the combination of two morphemes, the present tense with an abstract verbal morpheme often labeled "*woll*"; *would* is the corresponding form *PAST+woll* (Abusch 1988, 1997, 1998; Ogihara 1996; Condoravdi 2002; Kaufmann 2005b). Condoravdi and Kaufmann give *woll* the semantics of a modal; Ogihara and Abusch treat *woll* as a temporal operator, though not as a tense in the most narrow sense.

Enç (1996) gives two arguments that *will* is not a future tense. The first concerns sequence of tense, her (18)–(19):

(302)   (a)   Mary said that she was tired.
        (b)   Mary will say that she will be tired.

Enç notes that (302b) lacks the simultaneous reading. It is true if Mary says tomorrow "I will be tired," but not if she says tomorrow "I am tired." In other words, the embedded *will* cannot lack a semantic effect. It is a real, semantically active *will*. In this regard, it is different from the past tense in (302a), for which a simultaneous reading is possible. This difference makes sense if *will* is not a tense at all, but rather a modal.

Enç's second comparison concerns what happens when *will* takes scope over a present tense (her (22), (24)):

(303)   (a)   John said that Mary is upset.
        (b)   John will say that Mary is upset.

While there is controversy surrounding the exact meaning of (303a), it is clear that the present tense of *is* is linked to the speech time. (It may be linked to the past time at which John said what he said as well, a point which has led to this interpretation being called the DOUBLE ACCESS READING, cf. Enç 1987; Abusch 1988; Ogihara 1995.) Whatever the details of the meaning of this example are, the crucial point is that, in some respect, the time at which Mary is said to be upset includes the speaker's now.

Example (303b) is crucially different from (303a). The present tense of *is* in this case need not be linked to the speech time. The sentence has a simultaneous reading where, if the sentence is true, John claims at some future time *t* that Mary is upset at *t*, and John makes no claim about Mary's mood now, the time at which the whole sentence is uttered. Thus, this context presents another way in which *will* is different from the past tense.

Enç's two comparisons show clearly that *will* does not contain a future tense which behaves similarly to the past tense, and can be taken to prove that it is not a tense at all, provided that we mean by "tense" something which is rather like the past in morphosyntactic and/or semantic terms.[6] But her conclusion that it is therefore a modal is less well supported. Obviously *will* is similar to modals morphosyntactically, but are the semantic similarities sufficient to classify it as a modal? Enç offers two reasons to believe that *will* is a modal: the observation that *will* has modal uses and the fact that other modals involve reference to future situations.

Descriptive linguists such as Palmer (1986) as well as theoreticians like Enç point out that sentences containing *will* frequently express meaning other than simple futurity.

(304) (a) Pat will be sleeping now. (Enç 1996, (8))
       (b) Certain drugs will improve the condition. (Palmer 1986: 216)

Example (304a) has a present epistemic meaning, while (304b) is generic. The first example is not about the future at all, and so it's obvious that *will* is not a tense here. The second may be a quantificational modal. On the basis of examples like these, most linguists agree that *will* is sometimes a modal, and that the real question is whether it is always a modal.

Enç's second argument that *will* is a modal is that the kind of future meaning it displays is characteristic of modals more generally. In particular, she points out that deontic *must* also requires future orientation. But, as we saw in Section 5.1.1, a number of factors contribute to whether a modal sentence has future orientation. Therefore, Enç's argument that *will* has future meaning because it is a modal is a bit too quick. We should see whether its future meaning works the same way as the future orientation of modals. Enç makes a direct comparison

---

[6] Another point that is sometimes used (e.g., by Kaufmann 2005b) to support the idea that *will* is not a future tense is the fact that the present tense can be used to talk about future situations, as in *John leaves tomorrow*, and is therefore classified as a non-past tense (e.g., Palmer 1986). If one makes the assumption that a tense system would not allow one form (the non-past) to express a superset of the meanings expressed by another (the putative future), one can conclude that *will* is not a future tense. I don't know of any explicit argument for the assumption, however.

between *will* and deontic *must*, but we have seen that priority *must* and *may* are special among the modals in requiring future orientation. Moreover, their future orientation may be explained by their performative meaning (see Section 4.3.3). If *will* is a modal, it's a dynamic one, and so it would not have the same type of performativity as *must/may*. In looking at other modals, we have seen that both aspectual class and the choice of modal base help determine whether a modal sentence gets a future interpretation. The theory of Werner (2006) explains in a neat way why *will* has future meaning on the basis of the assumption that it employs a temporally restricted totally realistic modal base. While Werner's theory has some problems, if the long-standing idea that the future meaning of *will* follows from its status as a modal can be given a theoretical explanation, it is likely to be in terms of ideas similar to his.

One recent work, Kissine (2008), argues that *will* is never a modal. Instead, it has a purely temporal meaning. He adopts the temporal semantics of Abusch (1998) (very similar to the temporal part of Condoravdi 2002), according to which *will* extends the evaluation time from the some interval t to $[t, \_)$. If t is the speech time, then $[t, \_)$ is the present plus all moments in the future. Such a meaning is compatible with (304a), and, in combination with *now*, the sentence is equivalent to the corresponding present tense form.

Kissine presents an argument that *will* is not a modal operator, based on examples like the following (his 10)):

(305)   ?It is not the case that Mary will come and (for all that we know) it is possible$_{[epistemic]}$ that Mary will come.

If *will* is a modal operator more or less of the kind represented in modal logic, then this sentence has the form $\neg\Box p \wedge \Diamond\Box p$. (There's a corresponding point to be made with $\Box p \wedge \Diamond\Box\neg p$.) Kissine assumes that such sentences are odd because they are semantically inconsistent, and that their inconsistency is to be derived within a version of classical modal logic (in particular, modal logic augmented by Kratzer's view of context dependency, but without the ordering source). Within such a framework, the only way to derive inconsistency is to make assumptions about the modal operators according to which $\Box p \Leftrightarrow \Diamond\Box p$ (where $\Box$ represents *will* and $\Diamond$ represents *possible*). But this is an implausible result. (306a–b) are not equivalent:

(306)   (a)   Ben will win.
        (b)   It's possible that Ben will win.

Kissine's argument depends on the key assumption that, if *will* is a modal, we must explain the oddness of (305) in terms of the particular modal meaning attributed to *will*. However, we might take a different approach. Kissine sees example (305) as being of the form $\neg\Box p \wedge \Diamond\Box p$, but, replacing $\Box p$ by ϕ, it is also of the form $\neg\phi \wedge \Diamond\phi$. As we saw in Section 3.2, sentences of that form provided one of the primary motivations for a dynamic analysis of epistemic modality: $\neg\phi \wedge \Diamond\phi$ is inconsistent in dynamic semantics. Moreover, we even saw that a pragmatic explanation of the oddness of these examples is possible (though Yalcin 2007a argues against a pragmatic explanation on the grounds that the inconsistency is manifest in certain embedded structures). Thus, Kissine has not yet proven that *will* is not a modal.

Kissine admits that many sentences with *will* have modality as part of their meaning, and argues that this modality is due to a covert epistemic modal which is present in all root sentences which do not already contain an overt modal. In order for Kissine's analysis of the modality in sentences containing *will* to be plausible, he must motivate the presence of a covert modal in simple examples like (307) as well:[7]

(307)   (a)   Pat is asleep
        (b) $\Box(\text{PRES}(\text{Pat asleep}))$

Kissine adopts Stalnaker's theory of assertion, according to which under normal circumstances the proposition expressed by a declarative sentence is added to the common ground. In terms of this theory, what he shows is that, under the right assumptions about the nature of the covert modal, one can maintain that (307) contains such a modal. In particular, if the modal has the right properties, adding (307b) to the common ground will amount to adding the non-modal proposition (that Pat is asleep) to the common ground. However, he doesn't raise any considerations which show that modality is needed in the analysis of (307a). That is, we'd do just as well with the simple view that (307a) adds the proposition that Pat is asleep (with no modality) to the

---

[7] Though it is not his point, Kissine's analysis suggests another explanation for why sentences which appear to be of the form $\neg\phi \wedge \Diamond\phi$ are unacceptable. According to Kissine, there are no such sentences; any sentence which appears to have that form is actually $\Box\neg\phi \wedge \Diamond\phi$, where the $\Box$ is covert. In order to extend this approach to Yalcin's (2007a) data (see Section 3.2.3), he would have to propose that certain embedded clauses also have the covert modal. Note also that if there is a covert epistemic modal, as Kissine proposes, this explains why (305) is unacceptable independently of whether *will* is a modal: it would be of the form $\Box\neg p \wedge \Diamond p$. Thus, his second claim (that there is a covert modal) undermines the argument for his first claim (that *will* is not a modal).

common ground. Overall, then, Kissine's analysis requires one to accept an analysis of simple declarative sentences which is more complex than otherwise required in order to account for the modality present in sentences with *will*. It would be simpler to attribute the modality in examples with *will* to *will* itself, provided that such an analysis is not ruled out on other grounds. In other words, unless Kissine provides a convincing argument that *will* is not modal, there is no reason to accept the existence of the null modal in sentences with *will*.

The present

Many simple present tense sentences seem to involve a modal meaning.[8] Such sentences often convey generic meaning, as in (308a–b).

(308)　(a)　Dogs bark.
　　　　(b)　John smokes.

As we saw in Section 4.4, genericity involves modality; most scholars assume that generic meaning is introduced by a separate operator *GEN*, and is not incorporated into the present tense itself. The literature on genericity is complex and extensive, and so beyond the scope of this book. See Carlson and Pelletier (1995) for a good introduction and several important papers.

　　We also find modality in cases in which the present tense describes a future situation (from Kaufmann 2005b, (27)):

(309)　The plane leaves at 4 p.m.

Not only is this sentence about the future, a fact which may in and of itself imply a kind of modality; it moreover implies that his leaving is somehow determined or scheduled as of the speech time, and this type of meaning seems clearly modal (see Dowty 1979; Kaufmann 2005b, and references therein). For example, Kaufmann attributes the settledness of (309) to an unpronounced necessity modal using a doxastic or metaphysical (historical) modal base.[9] While I don't believe that (309) and similar examples can be analyzed just in terms of metaphysical and

---

[8] I will avoid discussing the temporal semantics of the so-called present tense here. Enç (1990) argues that English has no present tense (what we call "present tense" is the absence of tense), while other authors think that it does and enforces either a present or non-past meaning. The references cited in the text and Kuhn and Portner (2002) provide points of entry into the relevant literature.

[9] In fact, Kaufmann proposes that all sentences with either the simple present or past tense involve such a modal operator, but the arguments for this position are too complex to go into here.

doxastic modal bases (in this case, it need not be metaphysically determined that the plane leaves at 4 p.m., nor does the speaker have to fully
believe that it will), it's likely that a similar analysis involving ordering
sources and a wider range of conversational backgrounds would work.
It's plausible that (309) uses an epistemic modal base and a stereotypical
("normal circumstances") ordering source.

The past

Simple past tense sentence can be generic, as in (310), and in this respect
they can involve modality just as present tense sentences can.

(310)   A Tyrannosaur had small front limbs.

In many languages, a past tense can also be used to indicate the
"unreality" (Palmer 1986) of a situation. Consider the following data:

(311)   (a)   Suppose Jill knew French.
        (b)   Suppose Jill knows French.

(312)   An iχ_1e pari   to   siropi θa   iχ_1e $\gamma_1$ini   kala.
        if   had   taken the syrup FUT   had   become   better
        'If he had taken the syrup, he would have gotten better.'

$\hfill$ (Iatridou 2000, (5))

There is a difference between the meanings of (311a–b); the former
suggests that Jill probably does not know French, while the latter is
neutral on this point. The past tense can be used to indicate unreality
in conditionals as well, as in (312); such examples are often known as
"counterfactuals" or "subjunctive conditionals" in the philosophical
literature. The precise nature of the unreality conveyed by the past
tense in these examples is the subject of much debate. The concept of
counterfactual is clearly too strong, since (312) can be true in a situation
in which he did take the syrup.[10]

Here we are examining the relationship between tense and modality,
and so a key question is how the past tense comes to be associated with
unreality. As pointed out by Iatridou (2000), the association between
past and unreality is too common cross-linguistically to be considered

---

[10] This point was made early on by Anderson (1951). On the nature of the "counterfactuality" or "unreality" of this type of conditionals, see among many others Anderson
(1951); Lewis (1973); Stalnaker (1975); Karttunen and Peters (1975); Kratzer (1979); Portner
(1992, 1997); von Fintel (2001); Ippolito (2003); Veltman (2005); Gillies (2007a); Asher
and McCready (2007); Arregui (2007). We'll return to the semantics of conditionals in
Section 5.2 below.

an accident. Thus, it won't do to say that the apparent past tense on
knew in (311a) is simply a subjunctive with no connection to the past.
Isard (1974), Lyons (1977), and in much more theoretical detail Iatridou
(2000) develop analyses of the past which associate it with a uniform
abstract semantics which can be made concrete in terms of either
temporal or modal concepts. Thus, for Iatridou, the past indicates the
schema "the topic x excludes the x which is, as far as we know, the
x of the speaker," where x can be "time" or "world." If we let x be
instantiated by "time," this means that the time we are talking about
excludes the time which is, as far as we know, the utterance time.
Iatridou assumes that one needs a modal to talk about the future, and
so saying that we're talking about a time separate from (the time which
is, as far as we know,) the utterance time implies that we're talking
about the past. If x is instantiated as "world," the schema implies that
we are talking about a world which excludes the world which is, as far
as we know, the actual world. Iatridou takes this condition to correctly
describe the kind of unreality conveyed by the past tense.

Iatridou's analysis expresses an attractive intuition. Though on some
points the semantic features of her proposal are unclear, it is likely that
with some effort her proposal could be made semantically precise. Such
an analysis would go a long way towards providing an explanatory theory of the relationship between the temporal and unreality meanings of
the past tense.

5.1.3 The modality of aspect

Modality is an important component of the meanings of two of the
most well-studied aspectual constructions, the progressive and the perfect. In this section, we'll briefly look at each.

The progressive

The most famous semantic analysis relating aspect to modality is
Dowty's (1977, 1979) treatment of the progressive. His proposal amounts
to the claim that the progressive is a kind of modal with a unique
accessibility relation and some temporal meaning mixed in. The spe-
cial accessibility relation is INR; it relates a pair $\langle i, w \rangle$ consisting of
an interval of time and a world to a world. A world w' is accessible
from $\langle i, w \rangle$ (i.e., $w' \in \text{INR}(\langle i, w \rangle)$) iff everything which is going on in
w during i continues as it normally would in w'. We say that w' is
an INERTIA WORLD with respect to $\langle i, w \rangle$. INR is similar to a historical

accessiblity relation (in fact, Dowty's approach is naturally expressed in terms of branching time); it makes accessible a subset of the historically accessible worlds, those where things continue normally.

Based on INR, Dowty proposes the following meaning for the progressive operator PROG:

(313)   PROG(ϕ) is true at a pair of an interval and a world $\langle i, w \rangle$ iff for some interval i' which includes i as a non-final subinterval and for all inertia worlds $w' \in \text{INR}(\langle i, w \rangle)$, ϕ is true at $\langle i', w' \rangle$.

Let's take an example:

(314)   Mary was eating the apple.

According to Dowty, this sentence would be true at an interval i and world w iff in all of the worlds where what was going on in w during i continued as it normally would, Mary eats the apple during i or some longer interval starting with i. That is to say, in every inertia world, she eventually eats the apple.

Several problems have been noted for Dowty's analysis, and the most important one has to do with the fact that it predicts (314) to be false if some course of events independent of Mary and her apple is in progress which is likely to interrupt her snack. Suppose that while Mary took her first bite of the apple, a meteor was falling directly towards her head. It hits before she can finish the apple, knocking her unconscious or worse. In this case, (315) is intuitively true, but Dowty's analysis predicts that it is false:

(315)   Mary was eating the apple when the meteor struck her in the head.

The reason Dowty's analysis predicts (315) to be false is that (313) makes INR a function only of i and w. In w, the meteor is falling during i directly towards Mary, and so we'd have to say that, in the normal course of events, it prevents her from finishing the apple. Therefore, it's not the case that in all inertia worlds, she finishes the apple.

This type of problem provides much of the motivation for the other major theories of the progressive based on modality, those of Landman (1992), Bonomi (1997), and Portner (1998). (See these papers for references to other important work on the progressive.) These more recent modal theories share one characteristic difference from Dowty's analysis: they claim that the semantics of the progressive is crucially dependent on events. In particular, the set of possible worlds which play a role in the progressive's meaning is determined with reference to the event described by the sentence. Roughly speaking, in the case of (314)–(315), we don't want to consider what's going on across the entire world w during i; in particular, we don't want to consider the meteor. Rather, we want to focus only on the eating event itself.

The various newer modal theories of the progressive differ in precisely how they take events into account. This is not the place to explore the semantics of the progressive in great detail, so suffice it to say that of the three analyses mentioned above, Portner (1998) treats the progressive more like an ordinary modal than the others, Landman (1992) treats it less like an ordinary modals than the others, and Bonomi (1997) is somewhere in the middle. An evaluation of these theories depends on how well they allow us to explain a complex array of facts, and so one should look at the papers cited for more details. One issue which has not been explored is the relationship between the temporal meaning of the progressive and the temporal meanings of sentential modals. As we saw in Section 5.1.1, modal sentences can indicate present or future temporal meaning depending on such factors as the aspectual class of the clause under the scope of the modal and the type of accessibility relation. If the meaning of the progressive really is modal in nature, we might expect that the progressive can have future meaning under the same circumstances as modals.

In many languages, the form which expresses progressive meaning has other meanings as well. In English, the progressive seems to have a future interpretation in (316):

(316)   Mary is leaving tomorrow.

The semantics of this futurate progressive is discussed by Dowty (1979), among others. In Romance, the imperfect verbal form can be used with a habitual sense as well as a progressive sense. A number of analyses have extended the modal analysis of the progressive to these other meanings which are expressed by the same form, for example Dowty (1979), Cipria and Roberts (2000) and Ferreira (2004, 2005). To the extent that these analyses are successful, it supports the modal analysis of the progressive, and more generally the idea that aspect is closely connected to modality.

The perfect

Several semantic anlysis of the present perfect have proposed that it involves modal meaning as well. For example, Pancheva discusses the perfect in Bulgarian (the citation is Izvorski 1997), claiming that examples like the following incorporate an epistemic modal (her (13)):

(317)   Ivan izpil   vsičkoto vino včera.
         Ivan drunk all-the   wine yesterday
         'I apparently drank all the wine yesterday.'

This example has an inferential or evidential interpretation, as can be seen from the gloss; it incorporates perfect morphology, though in the third person the auxiliary is missing, a feature which distinguishes this "evidential perfect" from the regular, aspectual perfect. Pancheva proposes that the evidential meaning should be analyzed by means of an epistemic modal operator.

There is a clear typological link between the perfect and evidentiality (see, for example, Tatevosov 2001, 2003). However, the importance of this link is not clear. To begin with, in the case of Bulgarian there is some dispute as to whether the form in (317) does in fact have an evidential semantics.[11] Assuming that it does, the fact that it is a separate form from the aspectual perfect, though closely related, leaves open the question of whether the relation is grammatical or historical. Moreover, even if we decide that the perfect can express evidentiality, this still leaves open its relation to modality, since there is no consensus on the relation between evidentiality and epistemic modality. Some scholars have argued that the two are distinct, while others wish to reduce at least some cases of evidentiality to modality. Pancheva simply assumes that the evidential meaning is to be analyzed as a kind of modality, and does not consider alternative theories like those to be discussed in Section 5.3. (I should point out that these alternatives had not yet been developed at the time she wrote her paper.) Thus, at this point all we can say about the evidential perfect is that it has some kind of connection to modality, but at this point it is not clear how direct this link is.

Pancheva's work focused on a grammatically distinct evidential perfect, but later work on the English perfect suggests that modality contributes to its meaning as well. Katz (2003) argues that the perfect has a modal presupposition to the effect that an event of the same kind can occur again. For example, he says that (318) presupposes that Mary might visit New York again.

---

[11] See Friedman (1986) for discussion. Friedman argues against the traditional view that this form is evidential. However, he is considering the hypothesis that it is a reportative evidential, not a type of inferential evidential as discussed by Pancheva.

(318)   Mary has visited New York.

Portner (2003) also argues that the perfect conveys a modal pre-supposition, but according to him the presupposition expresses the relevance of the fact that Mary visited New York to the conversation. The notion of relevance is explicated in terms of the concept of a Question under Discussion developed by Ginzburg (1995a,b) and Roberts (1996). For example, suppose that we are interested in answering the question "Who knows of a good hotel in New York?" In such a context, the fact that Mary has visited New York, in combination with other background assumptions (e.g., Mary always does careful research on hotel options before making a reservation, Mary stayed in a hotel when she went to New York), entails an answer to this question, namely, that Mary knows of a good hotel. Portner formalizes this relationship as part of the meaning of the perfect. More precisely, he proposes that the conversational background consisting of the common ground plus the fact reported by (318) (i.e., that Mary visited New York) makes true a sentence of the form $\Box\phi$, where ϕ is an answer to the Question under Discussion (here $\Box\phi \approx$ "Mary must know of a good hotel in New York").

There seems to be a good reason to believe that there is some modal meaning expressed by the perfect, though the three works discussed above differ substantially in the kind of modality they propose to be relevant. In addition, of course, the perfect has some temporal meaning; for example, in (318), the event of Mary visiting New York must be in the past. A full understanding of the perfect requires an analysis of this pastness, quite a difficult topic in its own right (see Portner 2003 for discussion and references).

5.2 Conditionals

5.2.1 The relation between conditionals and modality

Conditionals are a complex topic and there is an enormous literature studying them. For this reason, even though semantics of conditionals is deeply tied up with modality, I will not be able to study them in any depth. Rather, I will confine myself to pointing out the main issues concerning the relationship between conditionals and modality, along the way pointing out some references which can help the reader begin further explorations.

First, some preliminaries. In a conditional sentence of the form *If A, B* or *B, if A*, the sentence *A* is known as the ANTECEDENT and *B* the CONSEQUENT. Linguistics often talk about "the *if* clause," and this term might refer either to *A* by itself or *if A*. The term "*if* clause" tends to suggest a compositional theory where clauses introduced by *if* are given a common analysis across all contexts in which they occur, while the term "antecedent" tends to suggest a logic-based analysis on which the entire conditional is analyzed as a single construction. (Similar points hold for "the *then* clause" and "consequent".) Philosophers' discussions of conditionals often focus on examples which do not contain any adverbial, generic, or modal operator in the *then* clause other than, perhaps, *will* or *would*. Such conditionals are often classified into two groups, the INDICATIVE conditionals and the SUBJUNCTIVE conditionals, based on the idea that they are more or less distinguished by the mood of the antecedent clause.[12] Counterfactuals constitute the central case of subjunctive conditionals, and future-oriented conditionals with *will* are sometimes included in the subjunctives as well (see references in Edgington 2006). In contrast, some philosophers and most linguists like to base their analyses on a wider range of examples (e.g., Lewis 1975; Farkas and Sugioka 1983).

Many semanticists adopt Kratzer's theory, discussed in Section 3.1, of the relationship between modals and conditional structures. Recall her claim that an *if* clause may serve to restrict a modal operator. More specifically, when a modal is the highest verbal element in the *then* clause, the *if* clause helps to determine the set of accessible worlds. In (319), the modal base f is epistemic and the ordering source g is doxastic.

(319)   If Mary is happy, John must be happy.

The *if* clauses creates the modal base $f^+(w) = f(w) \cup \{\{w : \text{Mary is happy in } w\}\}$. Example (319) is then equivalent to *John must be happy*, interpreted with respect to f^+ and g.

This view of the role of *if* clauses in modal sentences is representative of a more general function, namely, the ability of adverbial clauses, including *if* clauses, to restrict quantificational operators (Lewis 1975; Farkas and Sugioka 1983; von Fintel 1984). Recall the discussion of (270a), repeated below:

---

[12] We have seen in Section 5.1.2 that the use of past tense to indicate unreality, as is done in English, is common cross-linguistically, and it is a mistake to confuse this correlation of form and function with the subjunctive mood.

(320)   When Josephine flew an airplane, she always landed it safely.

The adverb of quantification *always* quantifies over situations. The *when* clause restricts the domain of quantification to situations in which Josephine flew an airplane, and so the sentence says (roughly) that, in all such situations, she landed it safely. Notice that *if* can replace *when* in this sentence without changing the meaning. That is, (321a) can mean the same thing as (320).

(321)   (a)   If Josephine flew an airplane, she always landed it safely.
         (b)   If Josephine flew an airplane, she landed it safely.

This shows that the *if* clause can restrict an adverb of quantification. Example (321b) lacks the adverb of quantification; it is ambiguous, but on one interpretation it has virtually the same meaning as (321a).

   (321b) has another meaning as well, one that is more apparent in the following variant:

(322)   If Josephine flew the airplane yesterday, she landed it safely.

Here the *if* clause does not restrict an adverb of quantification, and as a result the sentence is not about the frequency which which safe landings co-occur with Josephine flying an airplane. Nor does there appear to be a modal which the *if* clause can restrict. This type of example therefore poses the key puzzle for theories of conditionals. One idea, favored by most semanticists, is that this example contains a covert modal. Another, favored by many philosophers, is that it represents a separate conditional construction; if it is a separate constructions, it might be seen as altogether different from modals, or as somehow modal-like. In the remainder of this section, I will make some remarks about these three positions.

   The hypothesis that sentences like (322) contain a null modal has been made explicitly by Kratzer (1978, 1991a,b). Her idea is usually understood as involving a morpheme which is phonologically null but which otherwise has all of the syntactic and semantic properties of other modals. This way of understanding the modality of (322) involves a level of abstractness which some linguists find objectionable, but it fits naturally into the conception of the syntax/semantics interface which Kratzer, and many other formal semanticists, assume. One advantage of this way of looking at (322) is that it can be extended to other sentences which lack an explicit operator, but which are interpreted as if they had one: generic sentences (recall the discussion of the *Gen* operator from Section 4.4) and examples which can be seen as containing a null adverbial, e.g. (321b). An even more important methodological advantage is its strong predictive value. The covert modal approach implies that the semantics of a conditional can be accurately analyzed by providing a modal semantics for the hypothesized null element and then following the general principle that *if* clauses function as restrictors (in the case of modals, by affecting the modal base).

A closely related idea about conditionals has been developed within Discourse Representation Theory (DRT; Kamp 1981; Kamp and Reyle 1993). As we discussed in Section 4.4.1, DRT was was one of the original theories which developed Lewis's idea that an *if* clause can restrict an adverb of quantification. Later versions of DRT which incorporate modality agree with Kratzer (1977) in analyzing an *if* clause as an argument of a modal,[13] restricting the set of worlds over which it quantifies. In fact, Roberts (1989) has incorporated Kratzer's semantics for modality directly into DRT. The key difference between Kratzer's work and DRT consists not in how modal and conditional sentences are interpreted, but rather in the way the syntax/semantics interface is conceived.

DRT uses an intermediate level of representation between syntax and meaning, namely, the Discourse Representation Structure (DRS). In some cases when the syntax does not contain an operator, one is introduced into the DRS by the construction rules which derive DRSs from syntactic structures. For example, the conditional structure itself, in the absence of a syntactically present operator, results in the introduction of an operator in the DRS. This approach shares with Kratzer's the advantage of providing a uniform analysis of conditional sentences. Its syntax is less abstract than that of the approach which postulates a null modal, an advantage, but it is not committed to the prediction that sentences lacking overt operators will be interpreted in just the same manner as their analogues with them. In principle, a construction rule could introduce implicit modality into a position in the DRS which could never be occupied by an explicit modal. This difference may favor DRT in that the conditional (322) can have an implicitly modalized interpretation, but if one removes the *if* clause (*She landed it safely*), it cannot.[14] DRT will explain this fact by saying

---

[13] Frank (1996) argues that *if* clauses only restrict epistemic modals. The combination of *if* plus a priority modal is analyzed as a restricted covert epistemic modal with scope over the priority modal.

[14] As noted in Section 5.1.2, Kissine (2008) proposes that every sentence which does not contain an explicit modal has a covert epistemic modal instead. If he is right, then this argument in favor of the DRT approach may disappear.

that the conditional structure is needed to trigger the DRS construction rule which introduces the implicit modal.[15] But the issues are complex, and we cannot go into the relevant syntactic and semantic details here.

The idea that sentences with *if* involve a conditional construction with its own semantics is classically represented by the idea that *if p, then q* is to be represented by the material conditional as $(p \rightarrow q)$, as in standard logic. Recall that $(p \rightarrow q)$ is true in w iff p is false in w or q is true in w. While philosophers have defended this analysis (see e.g., von Wright 1957; Mackie 1973; Lewis 1973; Edwards 1974; Jackson 1979; and Veltman 1986), there is little chance that it can be correct for all conditional sentences. (Conditional sentences containing proportional adverbs of quantification like *usually* pose the most formidable puzzle, as pointed out by Lewis 1975 and subsequent work such as Heim's 1982.) Linguists are unwilling to accept the hypothesis that *if p then q* can have such radically different interpretations as material conditional vs. restricted adverbial quantification, and so this analysis seems to be a dead end for linguistic theory. That it may not be a dead end within philosophy only highlights the difference in goals between the two fields.

The work of Stalnaker (1968, 1975) and Lewis (1973, 1979a, 1981) on conditionals follows the classical analysis in treating conditionals as involving a two-place connective, but in contrast to the material conditional, the meaning they assign crucially relies on possible worlds semantics. The theories of Stalnaker and Lewis are alike in many ways, but they differ both in their scope and the semantic details. Stalnaker aims to give a unified analysis of indicative and subjunctive conditionals (what we called "real" and "unreal" conditionals in Section 5.1.2). Example (322), with an indicative verb form in the *if* clause, is an indicative conditional; (323) is a subjunctive conditional:

(323)   If Josephine had flown the airplane yesterday, she would have landed it safely.

Stalnaker analyzes (323) as follows: It is true in a world w iff the consequent is true in the world most similar to the actual world in which the antecedent is true. Let $f(p, w)$ pick out the world most similar to w in which p is true. Then:

---

[15] On the null modal theory, one would have to say either that the null modal must be licensed by an *if* clause, or that the non-conditional sentence has a difficult-to-perceive modal reading.

(324) $[\![\,(p > q)\,]\!]^w = 1$ iff $[\![\,q\,]\!]^{f(p,w)}$

For example, (323) is true if Josephine landed the plane safely in the world most similar to the actual world in which she flew the plane yesterday.

According to Stalnaker, the indicative conditional (322) has the same truth conditions as (323). At this point, two comments are sufficient: first, since the subjunctive (or "unreal") conditional in (323) contains an overt modal, from the perspective of the syntax/semantics interface it is natural to treat it as involving a restricted modal (assuming we accept that *if* clauses restrict modals in other cases[16]); and, second, as pointed out in Section 5.1.2, the precise pragmatic difference between the two kinds of conditionals is a controversial topic.

Lewis (1973, 1979a, 1981) develops an analysis of unreal conditionals similar to Stalnaker's. Instead of relying on f to pick out the world most similar to w in which q is true, Lewis proposes a function which associates each world with a more complex system of similarity among worlds. Specifically, he associates each world w with a "system of spheres of similarity." $S(w)$, the system of spheres associated with w, is a set of sets of worlds. These sets are centered on w, meaning that $\{w\} \in S(w)$, and nested, so that if A and B are both in $S(w)$, either $A \subseteq B$ or $B \subseteq A$. Each sphere includes all those worlds which are at least as similar to w as the member of the sphere which is the least similar to w. Hence, the smallest sphere is the set of worlds which are maximally similar to w, and hence it is just $\{w\}$, and if $A \subseteq B$, then all of those worlds in B which are not in A are less similar to w than any world in A.

Given the system of spheres, a subjunctive conditional with *would* has the following meaning:[17]

---

[16] Gillies (2007b) has recently argued that *if* clauses do not restrict epistemic modals, but rather that *if* is an operator with a semantics more or less of the kind proposed by Stalnaker and Lewis. However, he adds a component of meaning which allows it, as a side effect, to restrict an epistemic modal if one happens to be in its scope. It is difficult to see how to extend this analysis to non-epistemic modals and adverbs of quantification. It would make sense for Gillies to follow Frank's (1996) proposal that priority modals are never restricted, and that what appears to be a restricted priority modal is actually a restricted epistemic modal with scope over the priority modal, but this would still leave the puzzle of *if* in combination with a dynamic modal, generic operator, or adverbial quantifier.

[17] Lewis stated condition (ii) as in (ii)′, but I find it easier to grasp (ii):

(ii)′   for some $A \in S(w)$, $A \subseteq [\![\,(p \rightarrow q)\,]\!]^{w,S}$.

The smallest sphere in $S(w)$ with worlds
in which John not here.

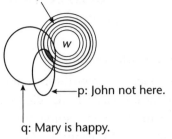

—p: John not here.

q: Mary is happy.

FIGURE 5.3. Lewis's semantics for counterfactuals

(325) $[\![\, p \mathbin{\square\!\!\rightarrow} q \,]\!]^{w,S} = 1$ iff

   (a)   for all $A \in S(w)$, $A \cap [\![\, p \,]\!]^{w,S} = \emptyset$ (there are no p-worlds
       in any sphere, the trivial case), or

   (b)   for some $A \in S(w)$, $A \cap [\![\, p \,]\!]^{w,S} \subseteq [\![\, q \,]\!]^{w,S}$.

An example:

(326)   If John were not here, Mary would be happy.

In Figure 5.3, the circles represent the system of spheres, so that the
smaller one of the circles is, the more similar the worlds in it are to w. In
w John is here and Mary is unhappy. The gray region represents those
worlds in which John is absent and which are as similar to w as possible
given this fact. Example (326) is true because all of these gray worlds
are ones in which Mary is happy. That is, we start with w and make the
minimal changes necessary to create a world in which John is absent.
There may be several ways to do this—he could be at a basketball game
or he could be at a movie—and all such alternatives are ones in which
Mary is happy.

Like the material conditional analysis, Stalnaker's and Lewis's theories fail to provide a uniform semantics for *if* clauses which can work
for all cases, including combinations of *if* with an adverb of quantification or a modal. Of course, it could be that different classes of
conditional sentences should receive different analyses, but the the idea
that sometimes they represent connectives ($\rightarrow$, $>$, or $\square\!\!\rightarrow$), while in
others they function to restrict operators, is not likely to be correct.
The difference is just too great to be bridged by plausible synchronic
or diachronic processes. Moreover, Kratzer has shown that with pro-
vision of implicit modals when necessary and the correct choice of
modal base and ordering source, the material conditional and Lewis's

counterfactual conditional come out as special cases of her theory. Therefore, the linguistic evidence favors the theory that *if* clauses always function to restrict an operator.

5.2.2 Conditionals and truth conditions

In section 4.2.1, we discussed the hypothesis that epistemic modals fail to contribute to truth conditions. A similar claim has been made about conditionals: according to this view, some conditional sentences do not have truth conditions (see, for example, Adams 1975; Gibbard 1981; Bennett 2003).[18] The claims are slightly different, however. Concerning modals, what is typically claimed is that epistemic modals themselves do not contribute to truth conditions, but this is compatible with the sentence containing the modal having truth conditions; indeed, scholars such as Westmoreland (1998) and Papafragou (2006) take the non-truth-conditional thesis to imply that a sentence with *must* is semantically equivalent to the corresponding sentence without *must*. In contrast, the corresponding claim about conditionals is that the entire conditional sentence lacks truth conditions.

Despite this difference, it is surprising that a link is rarely drawn between the issue of whether conditionals have truth conditions and the corresponding issues for epistemic modals. Suppose we assume that conditionals are to be analyzed in terms of the *if* clause restricting a (possibly null) modal operator. Then our assumptions about epistemic modals will obviously have important consequences for the analysis of conditionals.

1. If we think that epistemic modals do contribute to truth conditions, it will be hard to develop an analysis in which the conditional lacks truth conditions.
2. If we think that epistemic modals don't contribute to truth conditions, it seems that the *if* clause which restricts an epistemic modal would not either. From here, we must consider what it means to say that epistemic modals fail to contribute to truth conditions:
   (a) In the context of a view like Westmoreland's, which holds that the sentence with an epistemic modal has the same truth

---

[18] While this point is not made explicit, it appears that the claim is restricted to those conditionals which look like a separate construction, i.e., the philosophers' classic indicative and/or subjunctive conditionals. It probably is not meant to apply to conditionals which seem to involve restricted adverbs of quantification, conditional priority modality, generics. and the like.

conditions as the prejacent, this leads to the absurd conse-
quence that a conditional has the same truth conditions as its
(non-modalized) consequent.

(b) In the context of a view which holds that sentences containing
epistemic modals lack truth conditions altogether, we expect
that conditionals also lack truth conditions.

If we do not think that conditionals are to be analyzed in terms of the
if clause restricting a modal, the analyses of the two forms become
independent. It is possible to hold that one lacks truth conditions while
the other has them, for example. Nevertheless, it would be a responsible research strategy to consider them side by side. In this way, the
researcher could ensure that she identifies all of the consequences of
a given argument for or against the truth conditional analyses of both
constructions.

The analyses of conditionals and epistemic modals are also linked by
the fact that theories based on probability have been proposed for both
constructions. Ramsey (1965) suggested that the right way to assess your
belief in *if p, then q* is to hypothetically assume *p*, and then consider *q*.
For example, your level belief in (327) is based on the procedure of first
assuming that it's cold, and then seeing how strongly you believe that
it's snowing.

(327)   If it's cold, it's snowing.

Put in this way, Ramsey's point seems to be merely a description of
how we assess belief, but he goes further. He suggests that when you
hypothetically assume *p* and then consider *q*, you are determining a
probability of *q* given *p*, where "probability of *q* given *p*" is understood
as in probability theory. Once probability theory gets into the picture,
we are bound by its laws, in particular the one which relates lines (b)
and (c) in the following:

(328)   (a)   Your belief in *if p, then q* =
         (b)   The probability you assign to *q* given *p* =
         (c)   (The probability you assign to *p and q*)/(the probability
                you assign to *p*).

Thus, your level of belief in (327) should equal the probability you
assign to its being both cold and snowy, divided by the probability you
assign to its being cold.

Supposing we accept Ramsey's reasoning, the next question is what
it has to do with semantics. One might hope for the following picture:

p, *q*, and *if p, then q* all have truth conditions, and the semantics derives these truth conditions in the usual way (i.e., they express the proposition which is the set of possible worlds in which they are true). In addition, the probability assigned to each sentence is determined by the probability of the proposition it expresses. Thus, we would like an analysis of conditionals such that the probability of [[*if p, then q*]] = the probability of [[*p and q*]] /the probability of [[*p*]] . Unfortunately, Lewis (1976) proved that no such analysis of conditionals is possible. (Edgington 2001, 2006, provides a clear an intuitive explanation of this surprising result.)

Given that we cannot draw a direct link between the truth conditions of a conditional and the probability which an individual assigns to that conditional, perhaps we should forget about one or the other. On the one hand, we could say that people do not assign probabilities to conditionals (or at least do not do so in a coherent enough way to be worthy of precise analysis). Most scholars have assumed that this view is not tenable; Adams (1965, 1975) argues that in general assertability goes by subjective probability, that is a speaker can assert a sentence only if he assigns it a high enough probability. But, in that case, we must assign probabilities to conditionals, since they are sometimes assertable, and no better alternative has been discovered than the conditional probability in (328). On the other hand, we could give up on providing a truth-conditional semantics for conditionals: this is the non-truth-conditional view. If conditionals do not have a truth-conditional semantics, the next question is what kind of semantics they do have. Several authors suggest analyses based on probabilities (e.g., van Frassen 1976; McGee 1989; Jeffrey 1991; Stalnaker and Jeffrey 1994; Kaufmann 2005a and Yalcin 2007a).[19] As pointed out by Lewis, the non-truth-conditional view has difficulty explaining the interpretations of sentences which contain conditionals as constituents. Conditionals may be negated, conjoined, put into subordinate clauses, and so forth. We already have good analyses of these contexts, but the analyses are based on truth conditions. We should not give up on

---

[19] Except for Yalcin's, all of the papers cited here follow a hybrid approach which assigns truth conditions in cases in which the antecedent is true, and a probability otherwise. See Edgington (2006) for discussion and further references.

Yalcin's analysis departs from the standard truth-conditional view in two ways: it combines the "static" version of dynamic semantics outlined in Section 3.2.3 with probabilities. Thus far, his account has not been worked out in sufficient detail to make a careful evaluation feasible.

the idea that conditionals have truth conditions without a compelling reason.

Lewis (1976) and Jackson (1979) try to resolve this dilemma by claiming both that conditionals have truth conditions and that the assertability of a conditional is based on the conditional probability, as in (328). They do this by arguing that the pragmatic principles from which we determine what is assertable and what is not imply the importance of conditional probability. (In the postscript to Lewis 1976 which appears in Lewis 1986a, he accepts Jackson's account as better than his own earlier one.) If this approach is correct, we must then return to the question of which truth-conditional analysis is the best one. Lewis and Jackson both think that the material conditional analysis is acceptable, once a pragmatic theory involving probabilities is in place to explain assertability. But as pointed out above, the material conditional theory faces severe problems. Moreover, Edgington (2001, 2006) and Gibbard (1981) show that the material conditional theory fails to explain the meanings of complex sentences containing conditionals, even though such sentences were the main reason for holding onto truth conditions.

This discussion should make clear that the semantics of conditionals promises to remain a topic of continuous study for quite some time to come. What is important for our purposes here is the fact that, while we don't yet know what the relationship between modality and conditionals is, the problems we face in trying to understand them individually are very much alike. In both areas, we must contend with the issue of whether the semantics is truth conditional and the question of whether probability theory should be brought into the analysis. It seems likely that we will need to better understand both conditionals and epistemic modals before these general issues can be resolved. This remains true whether or not one accepts the more specific linguistic proposal that *if* clauses always serve to restrict an operator, and that sometimes this operator is a modal.

5.3 Modality, mood, and evidentiality

In this section, I will draw some links between modality and a range of other phenomena which are closely related to it. One way to categorize these topics is by their traditional labels: mood and evidentiality. I'm going to organize this section around these categories and their sub-categories. It also would have been possible to organize the discussion around the concepts of sub-sentential modality and discourse modality, discussed in Chapter 1. As we develop theories of mood, modality, and evidentiality, these categories are going to be crucial as well. For example, in trying to understand how the semantics of mood interacts with discourse, we'll need a precise theory of discourse meaning; and similarly with evidentiality. Of course, it should be the same theory of discourse meaning in both cases. Therefore, the notion of discourse modality is one we cannot ignore. That said, in this section I will not be delving into either sub-sentential modality or discourse modality in any detail. Rather, an exploration of these topics will have to await another venue, in particular *Mood* (Portner, forthcoming).

5.3.1 Mood

We can distinguish at least three types of phenomena to which the term "mood" has been applied: verbal mood, notional mood, and sentence mood. I will briefly discuss each in turn.

Verbal mood

The central members of this category are the indicative and subjunctive verb forms and the similar oppositions which go under different names in particular languages, for example "realis" and "irrealis" verbs; it can be taken to include other related forms as well (such as optatives and particles which have been assumed to mark subjunctive, as in Romanian). We may provisionally define this category as dependent sub-subsentential modality represented in the form of the verb.

Recent semantic theories of verbal mood have analyzed it as reflecting modal properties of the context in which the verb occurs. For example, in languages with a clear indicative/subjunctive opposition, a sentence translated as "It is possible that *S*" will typically have the tensed verb in *S* in the subjunctive:

(329)   Il est possible que je vienne.   (French)
        it is possible that I come.subjunctive
        'It is possible that I'll come.'

In (329), we can say that the subjunctive *vienne* is dependent on, is triggered by, or is licensed by *possible*. The subjunctive may be triggered in a wide range of contexts, and one of the main empirical challenges is to describe the range of triggers for each language and to understand the patterns of variation in the types of triggers relevant to different languages.

Not all triggers are as obviously modal as *possible*. For example, emotive constructions like *il est heureux que*... ('It is fortunate that...') require the subjunctive in French, as do interrogative and negative sentences in many cases (e.g., *Je ne pense pas que*..., 'I don't believe that...'). Subjunctive verbs can also occur in root clauses with a meaning similar to that which would be expressed by a sentence containing a main verb that would trigger the subjunctive. Despite the fact that the subjunctive is not dependent on sentential modals in these constructions, scholars have argued persuasively that they are all to be understood in terms of the same tools—quantification over possible worlds, ordering semantics, and the like—as sentential modality. But note that it is not simply the relevance of non-actual possible worlds which triggers the subjunctive; verbs translated as "believe" and "dream" often take the indicative, despite the fact that what is dreamed or believed may well not be true. As pointed out by Farkas (1985), this fact shows that it will not do to identify the indicative with "realis" meaning and the subjunctive with "irrealis." See Farkas (1985), Portner (1997, 1999), Quer (1998, 2001), Giorgi and Pianesi (1998), Giannakidou (1999, 2007), and Villalta (2000) for some of the more important theoretical work on the topic, as well as further references.

Notional mood

This term is used in a variety of ways, roughly to mean "something which is fundamentally the same as verbal mood, but which doesn't fit the strict definition." We can distinguish two main uses:

1. Forms with the function of verbal mood, but not expressed on the verb in the right way. This category includes, in particular, infinitives and dependent modals. Infinitives have a close connection to the subjunctive. For example, English uses infinitives in many contexts in which other languages would employ a subjunctive, for instance, in the complement of *want*, and French allows an infinitive in clauses which would otherwise require the subjunctive, when the subjects of the main and subordinate clause are the same (i.e., when a control structure is warranted). Some of the literature on mood, in particular Portner (1992, 1997), attempts to explain the relationship between infinitives and subjunctives.

   Harmonic modality was introduced in the discussion of (182)–(183) in Section 4.2.1. The phrase is used to describe cases in which two similar modal elements co-occur. (330) can be described as harmonic:

(330)   John can have the ability to swim.

The meaning of (330) involves two separate operators, and can be represented roughly as $\diamond\diamond\phi$. These two diamonds are alike, in that they both have ability interpretations. Such examples of harmonic modality are probably not very interesting for semantic theory.

In other cases, two harmonic modals seem to contribute only one instance of modality in the semantics, a situation which has been called "modal concord." A good example is (182a), repeated here:

(331)   Mary thinks it may rain.

As pointed out in Section 4.2.1, it may be possible to predict that the meaning of this sentence can be represented as $\diamond\phi$, with a single modal operator, on the basis of principles of modal logic. That is, two modal elements are really present, but their combination happens to be equivalent to a single operator.

There are other examples of modal concord where it seems better to think of the modal as dependent on another element, like mood, and not as introducing modal meaning of its own. Palmer's (183a), repeated here, is one example:

(332)   I pray that God may bless you.

Portner (1997) attempts to analyze this type of example within the semantic theory of verbal mood, essentially proposing that this kind of modal auxiliary, what I called a "dependent modal," is different from verbal mood only in its syntax. Geurts and Huitink (2006) give a similar analysis, though their approach only applies to cases where two modal elements, for example an auxiliary and an adverb, are clausemates, and they do not make the link to mood. Zeijlstra (2008) argues that modal concord is a syntactic, not a semantic, phenomenon.

2. An opposition between sentences in factual (sometimes "realis") and those in non-factual (sometimes "irrealis") mood. A sentence is in factual mood if it is meant to be interpreted as true or false in the actual world, and in the non-factual mood otherwise (Roberts 1989). (As pointed out above, it is sometimes suggested

that that verbal mood expresses this opposition, but this suggestion in wrong.) The material under the scope of a modal will typically be non-factual, in this sense, as will the antecedent of a conditional, an imperative clause, or a simple declarative sentence used to report on the content of a dream:

(333)   (a)   It might rain.
        (b)   Have a seat!
        (c)   I had a strange dream last night. My children were penguins.

The concept of (non)factual mood is especially important in the Discourse Representation Theory analysis of MODAL SUBORDINATION. Modal subordination is the phenomenon whereby a sentence is interpreted as part of the argument of a modal expression even though it does not occur under the syntactic scope of that modal. An example is (333c), where the second sentence is interpreted as under the scope of a "dream" operator: it means "I dreamed that my children were penguins." Here we say that the second sentence is modally subordinate to *dream*.

While (333c) relates to a subsentential modal operator (the noun *dream*), Roberts (1987, 1989, 2004) and others have extensively studied cases involving sentential modals, such as the following (Roberts 1989, (13)):

(334)   (a)   A thief might break into the house.
        (b)   He would take the silver.

(334b) means "If a thief were to break into the house, he would take the silver." The sentence (334b) is said to be modally subordinated to the first, more precisely to an *if* clause built out of the first.

The concept of non-factual mood proves useful in providing an analysis of modal subordination. Note that in (334), there is anaphora between the two sentences: *he* is linked to *a thief*, even though the indefinite *a thief* has scope under a modal. Such anaphora would not be possible if the second sentence were in the factual mood, as in (335):

(335)   (a)   A thief might break into the house.
        (b)   He is tall.

Though anaphora is possible here, it requires an interpretation on which *a thief* has scope over *might*. That is, *a thief* is not part of the constituent with nonfactual mood.

On the basis of examples like these, Roberts reaches the preliminary generalization that a pronoun in a factual mood sentence can never have an antecedent which is in a nonfactual mood sentence. (She refines this generalization in the course of her study, but this way of looking at the pattern will do for our purposes.) In the end, the concept of (non-)factual mood does not play a crucial role in Roberts theory. While it serves to organize the relevant data, in the end the analysis works in terms of more basic concepts of scope and discourse structure. Thus, we can see the DRT use of the concept of mood as essentially borrowing some associations of the core category, verbal mood, without requiring a real theory of mood over and above the broader theory of modality.

Sentence mood

This term is used for two related concepts which also go under the names CLAUSE TYPE (see Sadock and Zwicky 1985) and SENTENTIAL FORCE (Chierchia and McConnell-Ginet 1990; Portner 2004). The difference between clause type and sentential force is that the former relates to form, while the latter relates to meaning: The clause types include the categories of declarative, interrogative, and imperative, as well as other "minor types." The sentential forces are the conversational uses conventionally associated with these types, as in Table 5.1.[20] Some initial references are Stalnaker (1974, 1978), Heim (1982), and Groenendijk and Stokhof (1991) on assertion; Roberts (1996) and Ginzburg (1995a,b) on asking; and Lewis (1979b), Han (1998), and Portner (2004, 2007a) on requiring. There is also relevant work on the force of exclamatives (e.g., Zanuttini and Portner 2003).

Though "sentence mood" is sometimes used to mean clause type, it more often refers to sentential force (Reis 1999, 2003). It is important to distinguish sentence mood/sentential force from illocutionary force. Sentence mood is the conversational use conventionally associated with a particular grammatical category, the clause type. As such, all declaratives have the sentential force of assertion, even though in a particular situation a declarative may be used to ask a question or give an order. In contrast, the illocutionary force of a sentence is the type of

---

[20] I follow Portner (2004) in my labels for the forces.

TABLE 5.1. Major clause types and
associated sentential forces

Clause Type	Sentential Force
Declarative	Assertion
Interrogative	Asking
Imperative	Requiring

communicative act which the speaker intends on a particular occasion.
Thus, in the obvious context (336) has the sentential force of assertion
but the illocutionary force of asking:

(336)   I wonder if you can tell me the time.

Sentential force must be analyzed at the interfaces among syntax,
semantics, and pragmatics, while illocutionary force is a pragmatic phenomenon having to do with the speaker's communicative intentions,
analyzed in terms of speech act theory (e.g., Austin 1962; Searle 1969).

5.3.2 Evidentiality

Evidentiality may be defined in functional terms as the speaker's assessment of her grounds for saying what she does, and an evidential as a
grammatical form which expresses evidentiality. A clear example is the
reportative suffix *si* in Cuzco Quechua (from Faller 2006a):

(337)   Congresista-manta-s   haykuy-ta muna-n.
        congressman-ABL-REP enter-ACC want-3
        'He wants to be a Congressman. (=p)
        EVIDENTIAL: speaker was told that p

(The evidential is realized as *-s* here.) According to the functional definition, *I heard that*... is an evidential in (338), and this fact points to a
problem with employing the functional definition for the purposes of
developing a semantic theory of evidentiality.

(338)   I heard that it might rain tomorrow.

Across languages, many languages employ closed-class elements such
as affixes to indicate evidential meaning, and there are important typological properties of these morphemes (e.g., Willett 1988). It seems
that there is a grammatical category of evidentiality which must be
kept separate from the broader functional category. In recent years,

semanticists have focused a great deal of attention on evidentiality in this grammatical sense.

One of the main issues in the theory of evidentiality is its relationship to epistemic modality. As pointed out in Section 4.2.2, we must take care in considering this issue, as different scholars talk about it in different ways. Within the literature which attempts to give a precise analysis of evidentiality, we find prominently the following two views:

1. Evidentiality should be distinguished from epistemic modality because the two categories relate to quite different types of meaning. Here the main perspective is the one outlined in (a) below, but there are other versions as well, as in (b):
    (a) Evidentiality is a kind of performativity, in the sense defined in Chapter 4. Faller (2002) proposes that evidentials affect the speech act performed by a sentence (i.e., its sentential force), while epistemic modals contribute to truth conditions just as claimed by modal logic or a theory like Kratzer's. C. Davis et al. (2007) propose that evidential sentences both have the same non-modal meaning as a sentence without an evidential and affect the conversational context in other ways determined by the evidential. What's common to these approaches is the idea that evidentials operate outside of core, truth-conditional semantics.
    (b) Evidentiality is related to tense, aspect, and deixis. Chung (2007) argues that Korean evidentiality has to do with the ability of the speaker to perceive the evidence which provides the basis for what she says.
2. Evidentiality is a kind of epistemic modality; crucially, evidentials contribute to truth conditions. McCready and Ogata (2007) and Matthewson et al. (2008) take this view. However, both must argue for some non-standard ideas about the nature of epistemic modality in order to make their analyses work, so it might be better to understand these authors as proposing that evidentiality and epistemic modality together constitute a broader class, and that, within this class, what we call epistemic modals and evidentials are two common types.

Given that theoretical research on evidentiality is spread over such a wide range of languages, direct comparison of these theories is difficult. It could be that all of them are right when applied to the language they have been designed for—i.e., that Faller is right about Quechua, Chung about Korean, and so forth. In that case, we might reserve the term "evidential" for those forms which do not fall under an existing category like epistemic modality. This seems to be Faller's suggestion. Or we could return to the broadest functional definition of evidentiality, and come up with a new term (perhaps "evidence-based speech act modifiers") for any elements which fall under a theory like Faller's or Davis et al.'s.

5.4 Looking forward

In this chapter we have briefly investigated the relationship between modality and the areas of temporal semantics, conditionals, mood and evidentiality. What these have in common is that they have to do with the place of modality within intensional semantics. There is widespread, though not universal, agreement within semantics that we know, more or less, what modality is: it is quantification over possible worlds, or something to a very similar effect. Then in investigating the links between modality and these other topics, scholars consider which semantic ingredients must be used in addition to, or instead of, modality. For example, in the literature on temporal interpretation, we find different ideas about how the temporal semantics of modal sentences comes about—from the meaning of the modal, through the presence of a temporal operator, or via general principles of one sort or another.

There are of course links between modality and many other areas of semantics, too numerous to list here. But these will have to be left aside. I hope that, with the help of this book, no student of modality will find himself at a loss when he encounters a topic in linguistics which requires an understanding of modal semantics. We look forward to the results of those encounters.

Bibliography

Abusch, D. (1988). Sequence of tense, intensionality, and scope. In *The Proceedings of WCCFL 7*: 1–14. Stanford, CA: CSLI.

Abusch, D. (1997). Sequence of tense and temporal de re. *Linguistics and Philosophy*, 20: 1–50.

Abusch, D. (1998). Generalizing tense semantics for future contexts. In Rothstein, S. (ed.), *Events and Grammar*, pp. 13–33. Dordrecht: Kluwer Academic Publishers.

Adams, E. (1965). The logic of conditionals. *Inquiry*, 8: 166–97.

Adams, E. (1975). *The Logic of Conditionals*. Dordrecht: Reidel.

Anand, P. and Nevins, A. (2004). Shifty operators in changing contexts. In Watanabe, K. and Young, R. B. (eds), *Proceedings of Semantics and Linguistic Theory 14*. Ithaca, NY: CLC Publications, Cornell University Linguistics Department, pp. 20–37.

Anderson, A. R. (1951). A note on subjunctive and counterfactual conditionals. *Analysis*, 11: 35–8.

Anderson, A. R. (1956). The Formal Analysis of Normative Concepts. Technical Report 2, US Office of Naval Research.

Åqvist, L. (1967). Good samaritans, contrary-to-duty imperatives, and epistemic obligations. *Noûs*, 1(4) (1967): 361–79.

Åqvist, L. and Hoepelman, J. (1981). Some theorems about a 'tree' system of deontic tense logic. In Hilpinen, R. (ed.), *New Studies in Deontic Logic*, pp. 187–221. Dordrecht: Reidel.

Arregui, A. (2007). When aspect matters: the case of *would* conditionals. *Natural Language Semantics*, 15(3): 221–64.

Asher, N. (1986). Belief in discourse representation theory. *Journal of Philosophical Logic*, 15: 127–89.

Asher, N. and McCready, E. (2007). Were, would, might and a compositional account of counterfactuals. *Journal of Semantics*, 24: 93–129.

Athanasiadou, A., Canakis, C., and Cornillie, B. (eds) (2006). *Subjectification: Various Paths to Subjectivity*. Berlin and New York: Mouton de Gruyter.

Austin, J. (1962). *How to Do Things with Words*. New York: Oxford University Press.

Barbiers, S. (2001). Current issues in modality. In Barbiers, S., Beukema, F., and van der Wurff, W. (eds), *Modality and its Interaction with the Verbal System*, pp. 11–29. Amsterdam and Philadephia: Benjamins.

Barwise, J. and Perry, J. (1981). Situations and attitudes. *Journal of Philosophy*, 78: 668–91.

Bayert, A. (1958). La correction de la logique modale de second ordre S5 et adéquation de la logique modale de premier ordre S5. *Logique et Analyse*, 1: 99–121.

Beaver, D. (2001). *Presupposition and Assertion in Dynamic Semantics*. Studies in Logic, Language and Information. Stanford, CA: CSLI.

Bennett, J. (2003). *Philosophical Guide to Conditionals*. Oxford: Oxford University Press.

Bennett, J. and Partee, B. H. (1978). *Toward the Logic of Tense and Aspect in English*. Indiana University Linguistics Club; reprinted in Partee (2004).

van Bentham, J. and Liu, F. (2004). Dynamic logic of preference upgrade. *Journal of Applied Non-Classical Logics*, 14(2): 1–26.

Benveniste, E. (1971). Subjectivity in language. In *Problems in General Linguistics*, trans. Mary Elizabeth Meek, pp. 223–30. Coral Gables, FL. University of Miami Press; originally published in *Journal de psychologie* 55 (1958).

Berman, S. (1987). Situation-based semantics for adverbs of quantification. In Blevins, J. and Vainikka, A. (eds), *University of Massachusetts Occasional Papers in Linguistics 12*: Amherst: GLSA, pp. 45–68.

Bhatt, R. (1998). Obligation and possession. In Harley, H. (ed.), *Papers from the UPenn/MIT Roundtable on Argument Structure and Aspect*. Cambridge, MA: MITWPL, pp. 21–40.

Bhatt, R. (1999). *Covert Modality in Non-Finite Contexts*. PhD thesis, Philadelphia: University of Pennsylvania.

Blackburn, P., de Rejke, M., and Venema, Y. (2001). *Modal Logic*. Cambridge: Cambridge University Press.

Bonomi, A. (1997). The progressive and the structure of events. *Journal of Semantics*, 14: 173–205.

Brennan, V. (1993). *Root and Epistemic Modal Auxiliary Verbs*. PhD thesis, Amherst: University of Massachusetts.

Brisard, F. (2006). Logic, subjectivity, and the semantics/pragmatics distinction. In Athanasiadou, A., Canakis, C., and Cornillie, B. (eds), *Subjectification: Various Paths to Subjectivity*, pp. 41–74. Berlin: Mouton de Gruyter.

Brown, M. (1988). On the logic of ability. *Journal of Philosophical Logic*, 17(1): 1–26.

Bybee, J. and Fleischman, S. (eds) (1995a). *Modality in Grammar and Discourse*. Amsterdam: Benjamins.

Bybee, J. and Fleischman, S. (1995b). Modality in grammar and discourse: An introductory essay. In Bybee and Fleischman (1995a), pp. 1–14.

Bybee, J. L. and Pagliuca, W. (1985). Cross-linguistic comparison and the develoment of grammatical meaning. In Fisiak, J. (ed.), *Historical Semantics and Historical Word Formation*, pp. 60–3. Berlin: Mouton de Gruyter.

Bybee, J. L., Perkins, R. D., and Pagliuca, W. (1994). *The Evolution of Grammar: Tense, Aspect, and Modality in the Languages of the World*. Chicago, IL: University of Chicago Press.

Carlson, G. N. (1977). *Reference to Kinds in English*. PhD thesis, Amherst: University of Massachusetts.

Carlson, G. N. and Pelletier, F. J. (eds) (1995). *The Generic Book*. Chicago, IL: University of Chicago Press.

Chierchia, G. (1995a). *Dynamics of Meaning: Anaphora, Presupposition, and the Theory of Grammar*. Chicago, IL: University of Chicago Press.

Chierchia, G. (1995b). Individual-level predicates as inherent generics. In Carlson and Pelletier (1995), pp. 176–223.

Chierchia, G. and McConnell-Ginet, S. (1990). *Meaning and Grammar: An Introduction to Semantics*. Cambridge, MA: MIT Press.

Chisholm, R. M. (1963). Contrary-to-duty imperatives and deontic logic. *Analysis*, 24: 33–6.

Chung, K.-S. (2007). Spatial deictic tense and evidentials in Korean. *Natural Language Semantics*, 15: 187–219.

Cinque, G. (1999). *Adverbs and Functional Heads: A Crosslinguistic Perspective*. Oxford and New York: Oxford University Press.

Cipria, A. and Roberts, C. (2000). Spanish imperfecto and preterito: Truth conditions and aktionsart effects in a situation semantics. *Natural Language Semantics*, 8(4): 297–347.

Coates, J. (1983). *The Semantics of the Modal Auxiliaries*. London: Croom Helm.

Comrie, B. (1985). *Tense*. Cambridge: Cambridge University Press.

Condoravdi, C. (2002). Temporal interpretation of modals: Modals for the present and for the past. In Beaver, D., Kaufmann, S., Clark, B., and Casillas, L. (eds), *The Construction of Meaning*. Stanford, CA: CSLI, pp. 59–88.

Dancygier, B. and Sweetser, E. (2005). *Mental Spaces in Grammar*. Cambridge: Cambridge University Press.

Davidson, D. (1980). The logical form of action sentences (1967). In *Essays on Actions and Events*, ch. 6, pp. 196–248. Oxford: Clarendon Press.

Davis, C., Potts, C., and Speas, M. (2007). The pragmatic values of evidential sentences. In Gibson, M. and Friedman, T. (eds), *Proceedings of SALT 17*, Ithaca, NY: CLC Publications, forthcoming.

Davis, H., Matthewson, L., and Rullmann, H. (2007). A unified modal semantics for out-of-control in St'át'imcets. In Hogeweg, L., de Hoop, H., and Malchukov, A. (eds), *Proceedings of the TAMTAM Conference*, forthcoming.

de Haan, F. (2001). The relation between modality and evidentiality. *Linguistische Berichte*, 9: 201–16.

de Haan, F. (2006). Typological approaches to modality. In Frawley, W. (ed.), *The Expression of Modality*, pp. 27–69. Berlin: Mouton de Gruyter.

de Swart, H. (1993). *Adverbs of Quantification: A Generalized Quantifier Approach*, Outstanding Dissertations in Linguistics series. New York: Garland.

Dekker, P. (1993). Existential disclosure. *Linguistics and Philosophy*, 16(6): 561–87.

Dowty, D. (1977). Towards a semantic analysis of verb aspect and the English "imperfective" progressive. *Linguistics and Philosophy*, 1: 45–77.

Dowty, D. (1979). *Word Meaning and Montague Grammar*. Dordrecht: Reidel.

Drubig, H. (2001). On the syntactic form of epistemic modality. MS, Tübingen: University of Tübingen.

van Eck, J. (1982). A system of temporally relative modal and deontic predicate logic and its philosophical applications (2). *Logique et Analyse*, 25: 339–81.

Edgington, D. (1995). On conditionals. *Mind*, 104(414): 235–329.

Edgington, D. (2001). Conditionals. In Lou, G. (ed.), *The Blackwell Guide to Philosophical Logic*, pp. 385–414. Oxford: Blackwell Publishers.

Edgington, D. (2006). Conditionals. In Zalta, E. N. (ed.), *The Stanford Encyclopedia of Philosophy*. Available at: <http://plato.stanford.edu/archives/spr2006/entries/conditionals/>.

Edwards, J. S. (1974). A confusion about *if-then*. *Analysis*, 34(3): 84–90.

Egan, A., Hawthorne, J., and Weatherson, B. (2005). Epistemic modals in context. In Preyer, G. and Peter, G. (eds), *Contextualism in Philosophy: Knowledge, Meaning, and Truth*, pp. 131–70. Oxford: Oxford University Press.

Ehrich, V. (2001). Was *nicht müssen* und *nicht können* (nicht) bedeuten können: Zum Skopus der Negation bei den Modalverben des Deutschen. In Müller, R. and Reis, M. (eds), *Modalität und Modalverben im Deutschen*, *Linguistische Berichte Sonderhefte 9*. Hamburg: Helmut Buske Verlag, pp. 149–76.

van Eijck, J. (1996). Presuppositions and information updating. In de Swart, H., Kanazawa, M., and Piñon, C. (eds), *Quantifiers, Deduction, and Context*, pp. 87–110. Stanford: CSLI.

Enç, M. (1987). Anchoring conditions for tense. *Linguistic Inquiry*, 18(4): 633–57.

Enç, M. (1990). On the absence of the present tense morpheme in English. Unpublished manuscript. Madison: University of Wisconsin.

Enç, M. (1996). Tense and modality. In Lappin, S. (ed.), *The Handbook of Contemporary Semantic Theory*, Oxford: Blackwell, pp. 345–58.

Ernst, T. (2001). *The Syntax of Adjuncts*. New York: Cambridge University Press.

Faller, M. (2002). *Semantics and Pragmatics of Evidentials in Cuzco Quechua*. PhD thesis, Stanford, CA: Stanford University.

Faller, M. (2006a). The Cusco Quechua reportative evidential and rhetorical relations. MS., Manchester: University of Manchester.

Faller, M. (2006b). Evidentiality and epistemic modality at the semantics/pragmatics interface. Paper presented at the 2006 Workshop on Philosophy and Linguistics, Ann Arbor: University of Michigan.

Faller, M. (2006c). Evidentiality below and above speech acts. MS., Manchester: University of Manchester.

Farkas, D. (1985). *Intensional Descriptions and the Romance Subjunctive Mood.* New York: Garland.

Farkas, D. F. and Sugioka, Y. (1983). Restrictive *If/When* clauses. *Linguistics and Philosophy*, 6(2): 225–58.

Fauconnier, G. (1994). *Mental Spaces.* New York: Cambridge University Press.

Fauconnier, G. and Turner, M. (2002). *The Way We Think: Conceptual Blending and the Mind's Hidden Complexities.* New York: Basic Books.

Feldman, F. (1986). *Doing the Best We Can.* Dortrecht: Reidel.

Ferreira, M. (2004). Imperfectives and plurality. In Watanabe, K. and Young, R. B. (eds), *Proceedings of Semantics and Linguistic Theory 14*, Ithaca, NY: CLC Publications, Cornell University Linguistics Department, pp. 74–91.

Ferreira, M. (2005). *Event Quantification and Plurality.* PhD thesis, Cambridge, MA: Massachusetts Institute of Technology.

von Fintel, K. (1984). *Restrictions on Quantifier Domains.* PhD thesis, Amherst: University of Massachusetts.

von Fintel, K. (1995). A minimal theory of adverbial quantification, MS, Cambridge, MA: MIT.

von Fintel, K. (1999). NPI licensing, Strawson entailment, and context dependency. *Journal of Semantics*, 16(2): 97–148.

von Fintel, K. (2001). Counterfactuals in a dynamic context. In Kenstowicz, M. (ed.), *Ken Hale: A Life in Language*, pp. 123–52. Cambridge, MA: MIT Press.

von Fintel, K. (2003). Epistemic modals and conditionals revisited. Handout of a talk presented at the University of Massachusetts at Amherst, Dec. 12, 2003. <http://web.mit.edu/fintel/www/umass-handout.pdf>.

von Fintel, K. and Gillies, A. S. (2007a). "Might" made right. MS., Cambridge, MA: MIT and Ann Arbor: University of Michigan <http://mit.edu/fintel/mmr.pdf>.

von Fintel, K. and Gillies, A. S. (2007b). An opinionated guide to epistemic modality. In *Oxford Studies in Epistemology, Volume 2*, edited by Tamar Szabó Gendler and John Hawthorne, Oxford: Oxford University Press, pp. 36–62.

von Fintel, K. and Gillies, A. S. (2008). CIA leaks. *Philosophical Review*, 117(1): 77–98.

von Fintel, K. and Iatridou, S. (2006). *How to Say Ought in Foreign: The Composition of Weak Necessity Modals.* Paper presented at the 2006 Workshop on Philosophy and Linguistics, Ann Arbor: University of Michigan.

Frank, A. (1996). *Context Dependence in Modal Constructions.* PhD thesis, Stuttgart.

van Frassen, B. (1976). The logic of conditional obligation. *Journal of Philosophical Logic*, 1: 417–38.

Friedman, V. A. (1986). Evidentiality in the Balkans: Bulgarian, Macedonian, and Albanian. In Chafe, W. and Nichols, J. (eds), *Evidentiality: The Linguistic Coding of Epistemology*, pp. 168–87. Norwood, NJ: Ablex.

Gärdenfors, P. (2007). Representing actions and functional properties in conceptual spaces. In Ziemke, T. and Zlatev, J. and Frank, R. M. (eds), *Body, Language and Mind, Volume I: Embodiment*. Berlin: Mouton de Gruyter, pp. 167–95.

Garrett, E. (2000). *Evidentiality and Assertion in Tibetan*. PhD thesis, Los Angeles, CA: University of California at Los Angeles.

Gennari, S. (2003). Tense meaning and temporal interpretation. *Journal of Semantics*, 20(1): 35–71.

Gettier, E. (1963). Is justified true belief knowledge? *Analysis*, 23: 121–3.

Geurts, B. and Huitink, J. (2006). Modal concord. In Dekker, P. and Zeijlstra, H. (eds), *Proceedings of the ESSLLI Workshop Concord Phenomena at the Syntax-Semantics Interface, Malaga*, pp. 15–20. Available at <www.ru.nl/ncs/janneke/modalconcord-esslli06.pdf>.

Giannakidou, A. (1997). *The Landscape of Polarity Items*. PhD thesis, Groningen: University of Groningen.

Giannakidou, A. (1999). Affective dependencies. *Linguistics and Philosophy*, 22: 367–421.

Giannakidou, A. (2007). The dependency of the subjunctive revisited: temporal semantics and polarity. *Lingua (Special Issue on Mood, ed. J. Quer)*, forthcoming.

Gibbard, A. (1981). Two recent theories of conditionals. In Harper, W. L., Stalnaker, R., and Pearce, G. (eds), *Ifs*. Dordrecht: Reidel, pp. 211–47.

Gillies, A. S. (2007a). Counterfactual scorekeeping. *Linguistics and Philosophy*, 30: 329–60.

Gillies, A. S. (2007b). Iffiness. MS., Ann Arbor: University of Michigan. Available at <http://semanticsarchive.net/Archive/TI1OGVlY/iffiness.pdf>.

Ginzburg, J. (1995a). Resolving questions, part I. *Linguistics and Philosophy*, 18(5): 459–527.

Ginzburg, J. (1995b). Resolving questions, part II. *Linguistics and Philosophy*, 18(6): 567–609.

Giorgi, A. and Pianesi, F. (1998). *Tense and Aspect: From Semantics to Morphosyntax*. Oxford: Oxford University Press.

Gödel, K. (1933). Eine Interpretation des intuitionistischen Aussagenkalkülus. In *Ergebnisse eines mathematischen Kolloquiums*, vol. 4, pp. 34–40.

Grice, H. P. (1975). Logic and conversation. In Cole, P. and Morgan, J. (eds), *Speech Acts*, Syntax and Semantics 3. New York: Academic Press. pp. 41–58.

Groenendijk, J. and Stokhof, M. (1975). Modality and conversational information. *Theoretical Linguistics*, 2: 61–112.

Groenendijk, J. and Stokhof, M. (1990). Dynamic Montague grammar. In Kálman, L. and Pólos, L. (eds), *Papers from the Symposium on Logic and Language*. Budapest: Adakémiai Kiadó, pp. 3–48.

Groenendijk, J. and Stokhof, M. (1991). Dynamic predicate logic. *Linguistics and Philosophy*, 14: 39–100.

Groenendijk, J., Stockhof, M., and Veltman, F. (1996). Coreference and modality. In Lappin, S. (ed.), *The Handbook of Contemporary Semantic Theory*, pp. 179–213. Oxford: Blackwell.

Groenendijk, J., Stokhof, M., and Veltman, F. (1997). Coreference and modality in multi-speaker discourse. In Kamp, H. and Partee, B. (eds), *Context Dependence in the Analysis of Linguistic Meaning*, pp. 195–216. Stuttgart: I.M.S.

Hacking, I. (1967). Possibility. *The Philosophical Review*, 76(2): 143–68.

Hackl, M. (1998). On the semantics of "ability attributions". MS., Cambridge, MA: MIT Press.

Hacquard, V. (2006). *Aspects of Modality*. PhD thesis, Cambridge, MA: MIT.

Halliday, M. (1970). Functional diversity in language as seen from a consideration of modality and mood in English. *Foundations of Language*, 6: 322–61.

Han, C.-H. (1998). *The Structure and Interpretation of Imperatives: Mood and Force in Universal Grammar*. PhD thesis, Philadelphia: University of Pennsylvania.

Han, C.-H. (forthcoming). Imperatives. In Maienborn, C., von Heusinger, K., and Portner, P. (eds), *Semantics: An International Handbook of Natural Language Meaning*. Berlin: Mouton de Gruyter.

Hansson, B. (1969). An analysis of some deontic logics. *Noûs*, 3: 373–98.

Hansson, S. O. (1990). Preference-based deontic logic (pdl). *Journal of Philosophical Logic*, 19(1): 75–93.

Hansson, S. O. (2001). *Structures of Values and Norms*. Cambridge: Cambridge University Press.

Heim, I. (1982). *The Semantics of Definite and Indefinite Noun Phrases*. PhD thesis. Amherst: University of Massachusetts.

Heim, I. (1990). E-type pronouns and donkey anaphora. *Linguistics and Philosophy*, 13: 137–77.

Heim, I. (1982). Presupposition projection and the semantics of attitude verbs. *Journal of Semantics*, 9: 183–221.

Heim, I. and Kratzer, A. (1998). *Semantics in Generative Grammar*. Malden, MA, and Oxford: Blackwell.

Herzberger, H. (1979). Counterfactuals and consistency. *Journal of Philosophy*, 76: 83–8.

Hintikka, J. (1961). Modality and quantification. *Theoria*, 27: 110–28.

Hornstein, N. (1990). *As Time Goes By: Tense and Universal Grammar*. Cambridge, MA: MIT Press.

Horty, J. (2001). *Agency and Deontic Logic*. Oxford and New York: Oxford University Press.

Huddleston, R. and Pullum, G. K. (2002). *The Cambridge Grammar of the English Language*. Cambridge: Cambridge University Press.

Hughes, G. and Cresswell, M. (1996). *A New Introduction to Modal Logic*. London and New York: Routledge.

Iatridou, S. (2000). The grammatical ingredients of counterfactuality. *Linguistic Inquiry*, 31(2): 231–70.

Inman, M. V. (1993). *Semantics and Pragmatics of Colloquial Sinhala Nonvolitional Verbs*. PhD thesis, Stanford, CA: Stanford University.

Ippolito, M. (2003). Presuppositions and implicatures in counterfactuals. *Natural Language Semantics*, 11: 145–86.

Isard, S. (1974). What would you have done if.... *Theoretical Linguistics*, 1: 233–55.

Izvorski, R. (1997). The present perfect as an epistemic modal. In Lawson, A. and Cho, E. (eds), *The Proceedings of SALT VII*. Ithaca, NY: CLC Publications, Cornell University Linguistics Department, pp. 222–39.

Jackendoff, R. (1972). *Semantic Interpretation in Generative Grammar*. Cambridge, MA. MIT Press.

Jackson, F. (1979). On assertion and indicative conditionals. *The Philosophical Review*, 88(4): 565–89.

Jeffrey, R. (1991). Matter of fact conditionals. *Proceedings of the Aristotelian Society Supplementary Volume*, 65: 161–83.

Johnson, M. (1987). *The Body in the Mind. The Bodily Basis of Meaning, Imagination, and Reason*. Chicago, IL: University of Chicago Press.

Joos, M. (1964). *The English Verb: Form and Meanings*. Madison, WI: University of Wisconsin Press.

Kamp, H. (1979). The logic of historical necessity. Unpublished manuscript.

Kamp, H. (1981). A theory of truth and semantic representation. In Groenendijk, J., Janssen, T., and Stokhof, M. (eds), *Formal Methods in the Study of Language*, pp. 277–322. Amsterdam: Mathematical Centre.

Kamp, H. and Reyle, U. (1993). *From Discourse to Logic: Introduction to Modeltheoretic Semantics of Natural Language, Formal Logic and Discourse Representation Theory*. Dordrecht: Kluwer.

Kanger, S. (1957). *Provability in Logic*. Stockholm: Almqvist & Wiksell.

Kanger, S. (1971). New foundations for ethical theory. In Hilpinen, R. (ed.), *Deontic Logic: Introductory and Systematic Readings*, pp. 36–58. Dordrecht: D. Reidel.

Kaplan, D. (1989). Demonstratives: An essay on the semantics, logic, metaphysics, and epistemology of demonstratives and other indexicals. In

Almog, J., Perry, J., and Wettstein, H. (eds), *Themes from Kaplan*, pp. 48–614. New York: Oxford University Press.

Karttunen, L. (1972). Possible and must. In Kimball, J. (ed.), *Syntax and Semantics, volume 1*, pp. 1–20. New York: Academic Press.

Karttunen, L. and Peters, S. (1975). Conventional implicature. In Oh, C.-Y. and Dinneen, D. (eds), *Syntax and Semantics, vol. 11: Presupposition*, pp. 1–56. New York: Academic Press.

Katz, G. (2003). A modal account of the English present perfect puzzle. In Young, R. B. and Zhou, Y. (eds), *Proceedings of Semantics and Linguistic Theory 13*. Ithaca, NY: CLC Publications. pp. 141–61.

Kaufmann, S. (2005a). Conditional predictions. *Linguistics and Philosophy*, 28: 181–231.

Kaufmann, S. (2005b). Conditional truth and future reference. *Journal of Semantics*, 22(3): 231–80.

Kaufmann, S., Condoravdi, C., and Harizanov, V. (2006). Formal approaches to modality. In Frawley, W. (ed.), *The Expression of Modality*, pp. 71–106. Berlin and New York: Mouton de Gruyter.

Kennedy, C. (2007). Vagueness and grammar: the semantics of relative and absolute gradable adjectives. *Linguistics and Philosophy*, 30(1): 1–45.

Kenny, A. (1975). *Will, Freedom, and Power*. Dordrecht: D. Reidel.

Kenny, A. (1976). Human abilities and dynamic modalities. In Manninen, J. and Tuomela, R. (eds), *Essays on Explanation and Understanding: Studies in the Foundations of Humanities and Social Sciences*, pp. 209–32. Dordrecht: D. Reidel.

Kissine, M. (2008). Why *will* is not a modal. *Natural Language Semantics*, 16(2): 129–55.

Klein, E. (1980). A semantics for positive and comparative adjectives. *Linguistics and Philosophy*, 4(1): 1–45.

Klinedinst, N. (2007). Plurals, possibilities, and conjunctive disjunction. *UCL Working Papers in Linguistics*, 19: 261–84.

Kratzer, A. (1977). What "must" and "can" must and can mean. *Linguistics and Philosophy*, 1(1): 337–55.

Kratzer, A. (1978). *Semantik der Rede*. Königstein: Scriptor.

Kratzer, A. (1979). Conditional necessity and possibility. In Bäuerle, R., Egli, U., and von Stechow, A. (eds), *Semantics from Different Points of View*, pp. 117–47. Berlin: Springer.

Kratzer, A. (1981). The notional category of modality. In Eikmeyer, H.-J. and Rieser, H. (eds), *Words, Worlds, and Contexts*, pp. 38–74. Berlin: de Gruyter.

Kratzer, A. (1986). Conditionals. In Farley, A. M., Farley, P., and McCollough, K. E. (eds), *Papers from the Parasession on Pragmatics and Grammatical Theory*, pp. 115–135. Chicago, IL: Chicago Linguistics Society.

Kratzer, A. (1989). An investigation of the lumps of thought. *Linguistics and Philosophy*, 12: 607–53.

Kratzer, A. (1991a). Conditionals. In von Stechow, A. and Wunderlich, D. (eds), *Semantik/Semantics: An International Handbook of Contemporary Research*, pp. 651–6. Berlin: de Gruyter.

Kratzer, A. (1991b). Modality. In von Stechow, A. and Wunderlich, D. (eds), *Semantik/Semantics: An International Handbook of Contemporary Research*, pp. 639–50. Berlin: de Gruyter.

Kratzer, A. (1995). Stage-level and individual-level predicates. In Carlson, G. and Pelletier, F. (eds), *The Generic Book*, pp. 125–75. Chicago, IL: University of Chicago Press.

Kratzer, A. (spring 2007). Situations in natural language semantics. In Zalta, E. N. (ed.), *The Stanford Encyclopedia of Philosophy*, available at: <http://plato.stanford.edu/archives/spr2007/entries/situations-semantics/>.

Krifka, M. (2001). Quantifying into question acts. *Natural Language Semantics*, 9: 1–40.

Krifka, M., Pelletier, F. J., Carlson, G. N., ter Meulen, A., Link, G., and Chierchia, G. (1995). Genericity: an introduction. In Carlson, G. N. and Pelletier, F. J. (eds), *The Generic Book*, pp. 1–124. Chicago, IL: University of Chicago Press.

Kripke, S. (1959). A completeness theorem in modal logic. *The Journal of Symbolic Logic*, 24: 1–14.

Kripke, S. (1963). Semantical analysis of modal logic I, normal propositional calculi. *Zeitschrift für mathematische Logic und Grundlagen der Mathematik*, 9: 67–96.

Kuhn, S. and Portner, P. (2002). Tense and time. In Gabbay, D. and Guenthner, F. (eds), *Handbook of Philosophical Logic*, vol. VI, 2nd edn, pp. 277–346. Dordrecht: Reidel.

Lakoff, G. (1987). *Women, Fire and Dangerous Things: What Categories Reveal About the Mind*. Chicago, IL: University of Chicago Press.

Lakoff, G. and Johnson, M. (1980). *Metaphors We Live By*. Chicago, IL: University of Chicago Press.

Landman, F. (1992). The progressive. *Natural Language Semantics*, 1: 1–32.

Langacker, R. W. (1985). Observations and speculations on subjectivity. In Haiman, J. (ed.), *Iconicity in Syntax*, pp. 109–50. Amsterdam: Benjamins.

Langacker, R. W. (1990). Subjectification. *Cognitive Linguistics*, 1: 5–38.

Langacker, R. W. (1999). *Grammar and Conceptualization*. Berlin and New York: Mouton de Gruyter.

Langacker, R. W. (2006). Subjectification, grammaticalization, and conceptual archetypes. In Athanasiadou, A., Canakis, C., and Cornillie, B. (eds), *Subjectification: Various Paths to Subjectivity*. Berlin: Mouton de Gruyter, pp. 17–40.

Leech, G. (1971). *Meaning and the English Verb*. London: Longman.

Lewis, C. I. (1918). *A Survey of Symbolic Logic*. Berkeley, CA: University of California Press.

Lewis, D. K. (1972). *General Semantics*. Dordrecht: Reidel.

Lewis, D. K. (1973). *Counterfactuals*. Cambridge, MA: Harvard University Press.

Lewis, D. K. (1975). Adverbs of quantification. In Keenan, E. (ed.), *Formal Semantics of Natural Language*, pp. 3–15. Cambridge: Cambridge University Press.

Lewis, D. K. (1976). Probabilities of conditionals and conditional probabilities. *Philosophical Review*, 85: 297–315; reprinted in Lewis (1986a), 133–56.

Lewis, D. K. (1979a). Counterfactual dependence and time's arrow. *Noûs*, 13: 455–76.

Lewis, D. K. (1979b). A problem about permission. In Saarinen, E., Hilpinen, R., Niiniluoto, I., and Hintikka, M. P. (eds), *Essays in Honour of Jaakko Hintikka*, pp. 163–75. Dordrecht: Reidel.

Lewis, D. K. (1981). Ordering semantics and premise semantics for counterfactuals. *Journal of Philosophical Logic*, 10: 217–34.

Lewis, D. K. (1986a). *Philosophical Papers, vol. 2*. Oxford: Oxford University Press.

Lewis, D. K. (1986b). *The Plurality of Worlds*. Oxford: Blackwell.

Link, G. (1983). The logical analysis of plurals and mass terms: A lattice-theoretical approach. In Bauerle, R., Schwartze, C., and von Stechow, A. (eds), *Meaning, Use and Interpretation in Language*, pp. 302–23. Berlin: Mouton de Gruyter.

Lyons, J. (1977). *Semantics*. Cambridge: Cambridge University Press.

McClosky, J. (2006). Questions and questioning in a local English. In Zanuttini, R., Herburger, E., Campos, H., and Portner, P. (eds), *Negation, Tense, and Clausal Architecture: Cross-Linguistic Investigations*. Washington, DC: Georgetown University Press, pp. 87–126.

McCready, E. and Ogata, N. (2007). Evidentiality, modality and probability. *Linguistics and Philosophy*, 30(2): 147–206.

McDowell, J. (1987). *Assertion and Modality*. PhD thesis, Los Angeles: USC.

MacFarlane, J. (2006). Epistemic modals are assessment sensitive. Ms., University of California at Berkeley. To appear in a collection edited by Brian Weatherson and Andy Egan.

McGee, V. (1989). Conditional probabilities and compounds of conditionals. *Philosophical Review*, 98: 485–542.

Mackie, J. L. (1973). *Truth, Probability, and Paradox*. Oxford: Oxford University Press.

McNamara, P. (2006). Deontic logic. In Zalta, E. N. (ed.), *The Stanford Encyclopedia of Philosophy*, available at: <http://plato.stanford.edu/archives/spr2006/entries/logic-deontic/>.

Marrano, A. M. (1998). *The Syntax of Modality: A Comparative Study of Epistemic and Root Modal Verbs in Spanish and English*. PhD thesis, Washington, DC: Georgetown University.

Mastop, R. (2005). *What Can You Do: Imperative Mood in Semantic Theory*. PhD thesis, Amsterdam: University of Amsterdam.

Matthewson, L., Rullmann, H., and Davis, H. (2006). Evidentials are epistemic modals in St'át'imcets. In Masaru Kiyota, James J. Thompson, and Noriko Yamane-Tanaka (eds), *Papers for the International Conference on Salishan and Neighbouring Languages XLI, University of British Columbia Working Papers in Linguistics*, Volume 18, Vancouver: University of British Columbia, pp. 221–63.

Matthewson, L., Rullmann, H., and Davis, H. (2008). Evidentials as epistemic modals: Evidence from St'át'imcets. In Jeroen Van Craenebroeck and Johan Rooryck (eds), *The Linguistic Variation Yearbook*, 7. Amsterdam: John Benjamins, pp. 201–54.

Milsark, G. (1977). Toward an explanation of certain peculiarities of the existential construction in English. *Linguistic Analysis*, 3: 1–30.

Mitchell, J. (1986). *The Formal Semantics of Point of View*. PhD thesis, Amherst: University of Massachusetts.

Moltmann, F. (2005). Relative truth and the first person. Available at <http://semanticsarchive.net/Archive/mY2NGJhY/>.

Mondadori, F. (1978). Remarks on tense and mood: The perfect future. In Guenthner, F. and Rohrer, C. (eds), *Studies in Formal Semantics: Intensionality, Temporality, Negation*. Amsterdam: North Holland Press, pp. 223–48.

Montague, R. (1960). Logical necessity, physical necessity, ethics and quantifiers. *Inquiry*, 4: 259–69.

Montague, R. (1973). The proper treatment of quantification in ordinary English. In Hintikka, K., Moravcsik, J., and Suppes, P. (eds), *Approaches to Natural Language*, pp. 221–42. Dordrecht: Reidel.

Mortelmans, T. (2000). On the 'evidential' nature of the 'epistemic' use of the German modals müssen and sollen. In van der Auwera, J. and Dendale, P. (eds), *Modal Verbs in Germanic and Romance Languages, Belgian Journal of Linguistics*, pp. 131–48.

Narrog, H. (2005). Modality, mood, and change of modal meanings: a new perspective. *Cognitive Linguistics*, 16(4): 677–731.

Ninan, D. (2005). Two puzzles about deontic necessity. In Gajewski, J., Hacquard, V., Nickel, B., and Yalcin, S. (eds), *New Work on Modality*, vol. 51 of *MIT Working Papers in Linguistics*, pp. 149–78. Cambridge, MA: MITWPL.

Nuyts, J. (1993). Epistemic modal adverbs and adjectives and the layered representation of conceptual and linguistic structure. *Linguistics*, 31: 933–69.

Nuyts, J. (2001a). *Epistemic Modality, Language, and Conceptualization*. Amsterdam and Philadephia: John Benjamins.

Nuyts, J. (2001b). Subjectivity as an evidential dimension in epistemic modal expressions. *Journal of Pragmatics*, 33: 383–400.

Nuyts, J., Byloo, P., and Diepeveen, J. (2005). *On Deontic Modality, Directivity, and Mood: A Case Study of Dutch Mogen and Moeten*, vol. 110 of Antwerp Papers in Linguistics, Center for Grammar, Cognition, and Typology, University of Antwerp.

Ogihara, T. (1989). *Temporal Reference in English and Japanese*. PhD thesis, Austin, TX: University of Texas at Austin.

Ogihara, T. (1995). Double access sentences and references to states. *Natural Language Semantics*, 3: 177–210.

Ogihara, T. (1996). *Tense, Attitudes, and Scope*. Dordrecht and Boston: Reidel.

Palmer, F. (1986). *Mood and Modality*. Cambridge: Cambridge University Press.

Palmer, F. (1990). *Modality and the English Modals*. New York: Longman.

Palmer, F. (2001). *Mood and Modality*, 2nd edn. Cambridge: Cambridge University Press.

Papafragou, A. (2006). Epistemic modality and truth conditions. *Lingua*, 116: 1688–702.

Partee, B. H. (1988). Possible worlds in model-theoretic semantics: A linguistic perspective. In Allen, S. (ed.), *Possible Worlds in Humanities, Arts, and Sciences: Proceedings of Nobel Symposium 65*, pp. 93–123. Berlin and New York: Walter de Gruyter.

Partee, B. H. (2004). *Compositionality in Formal Semantics: Selected Papers by Barbara H. Partee*. Oxford: Blackwell.

Perkins, M. R. (1983). *Modal expressions in English*. Norwood, NJ: Ablex.

Perlmutter, D. (1971). *Deep and Surface Structure Constraints in Syntax*. New York: Holt, Rinehart and Winston.

Pollock, J. (1976). *Subjunctive Reasoning*. Dordrecht: Reidel.

Portner, P. (1992). *Situation Theory and the Semantics of Propositional Expressions*. PhD thesis, Amherst: University of Massachusetts.

Portner, P. (1997). The semantics of mood, complementation, and conversational force. *Natural Language Semantics*, 5: 167–212.

Portner, P. (1998). The progressive in modal semantics. *Language*, 74: 760–87.

Portner, P. (1999). The semantics of mood. *Glot International*, 4(1): 3–9.

Portner, P. (2003). The temporal semantics and modal pragmatics of the perfect. *Linguistics and Philosophy*, 26: 459–510.

Portner, P. (2004). The semantics of imperatives within a theory of clause types. In Watanabe, K. and Young, R. B. (eds), *Proceedings of Semantics and Linguistic Theory 14*. Ithaca, NY: CLC Publications, Cornell University Linguistics Department, pp. 235–52.

Portner, P. (2005). *What is Meaning? Fundamentals of Formal Semantics*. Oxford: Blackwell.

Portner, P. (2007a). Beyond the common ground: The semantics and pragmatics of epistemic modals. In Yoon, J.-Y. and Kim, K.-A. (eds), *The Perspectives of Linguistics in the 21st Century*. Seoul: Hankook Publishing Company.

Portner, P. (2007b). Imperatives and modals. *Natural Language Semantics*, 15(4): 351–83.

Portner, P. (forthcoming). *Mood*. Oxford: Oxford University Press.

Potts, C. (2005). *The Logic of Conventional Implicature*. Oxford: Oxford University Press.

Prior, A. N. (1958). Escapism: the 'logical basis of ethics'. In Melden, A. I. (ed.), *Essays in Moral Philosophy*, pp. 135–46. Seattle: University of Washington Press.

Quer, J. (1998). *Mood at the Interface*. The Hague: Holland Academic Graphics.

Quer, J. (2001). Interpreting mood. *Probus*, 13: 81–111.

Quirk, R., Greenbaum, S., Leech, G., and Svartvik, J. (1985). *A Comprehensive Grammar of the English Language*. London: Longman.

Ramsey, F. P. (1931; reprinted 1965). Truth and probability. In Braithwaite, R. B. (ed.), *Foundations of Mathematics and Other Logical Essays*. London: Routledge and Kegan Paul, pp. 156–98; also reprinted in Ramsey (1990).

Ramsey, F. P. (1990). *Philosophical Papers*, ed. D. H. Mellor. Cambridge: Cambridge University Press.

Reis, M. (1999). On sentence types in German: An enquiry into the relationship between grammar and pragmatics. *IJGLSA*, 4(2): 195–236.

Reis, M. (2003). On the form and interpretation of German wh-infinitives. *Journal of Germanic Linguistics*, 15(2): 155–201.

Roberts, C. (1987). *Modal Subordination, Anaphora, and Distributivity*. PhD thesis, Amherst: University of Massachusetts.

Roberts, C. (1989). Modal subordination and pronominal anaphora in discourse. *Linguistics and Philosophy*, 12: 683–721.

Roberts, C. (1996). Information structure in discourse: Towards an integrated formal theory of pragmatics. In Toon, J.-H. and Kathol, A. (eds), *Papers in Semantics*, OSU Working Papers in Linguistics, vol. 49, Department of Linguistics. Columbus: Ohio State University, Linguistics Department, pp. 91–136.

Roberts, C. (2004). Context in dynamic interpretation. In *Handbook of Pragmatics*, edited by Laurence Horn and Gregory Ward. Oxford and Malden, MA: Blackwell, pp. 197–220.

Roberts, I. (1985). Agreement parameters and the development of English auxiliaries. *Natural Language and Linguistic Theory*, 3: 21–58.

van Rooij, R. (2000). Permission to change. *Journal of Semantics*, 17: 119–45.

Ross, J. R. (1969). Auxiliaries as main verbs. In Todd, W. (ed.), *Studies in Philosophical Linguistics (Series 1)*. Evanston, IL: Great Expectations Press, pp. 77–102.

Rullmann, H., Matthewson, L., and Davis, H. (2007). Modals as distributive indefinites. *Natural Language Semantics*, forthcoming.

Sadock, J. M. and Zwicky, A. (1985). Speech act distinctions in syntax. In Shopen, T. (ed.), *Language Typology and Syntactic Description*, pp. 155–196. Cambridge: Cambridge University Press.

Schlenker, P. (2003). A plea for monsters. *Linguistics and Philosophy*, 26: 29–120.

Searle, J. (1969). *Speech Acts*. Cambridge: Cambridge University Press.

Sharvit, Y. (2004). Free indirect discourse and de re pronouns. In Watanabe, K. and Young, R. B. (eds), *Proceedings of Semantics and Linguistic Theory 14*. Ithaca, NY: CLC Publications, Cornell University Linguistics Department, pp. 305–22.

Stalnaker, R. (1968). A theory of conditionals. In Rescher, N. (ed.), *Studies in Logical Theory*, pp. 315–32. Oxford: Blackwell.

Stalnaker, R. (1974). Pragmatic presupposition. In Munitz, M. and Unger, P. (eds), *Semantics and Philosophy*, pp. 197–213. New York: New York University Press.

Stalnaker, R. (1975). Indicative conditionals. *Philosophia*, 5: 269–86.

Stalnaker, R. (1978). Assertion. In Cole, P. (ed.), *Syntax and Semantics 9: Pragmatics*, pp. 315–32. New York: Academic Press.

Stalnaker, R. (1987). *Inquiry*. Cambridge, MA: MIT Press.

Stalnaker, R. and Jeffrey, R. (1994). Conditionals as random variables. In Eells, E. and Skyrms, B. (eds), *Probability and Conditionals*. Cambridge: Cambridge University Press, pp. 31–46.

Stephenson, T. (2007). A parallel account of epistemic modals and predicates of personal taste. In Puig-Waldmüller, E. (ed.), *Proceedings of Sinn und Bedeutung*, pp. 583–97. Barcelona: Universitat Pompeu Fabra; MS, MIT.

Stone, M. (1994). The reference argument of epistemic *must*. In *The Proceedings of IWCS 1*: 181–90.

Swanson, E. (2006a). *Interactions with Context*. PhD thesis, Cambridge, MA: MIT.

Swanson, E. (2006b). Something "might" might mean. MS., Ann Arbor: University of Michigan.

Swanson, E. (2007). The language of subjective uncertainty. MS, Ann Arbor: University of Michigan; revised version of Ch. 2 of his 2006 MIT dissertation.

Sweetser, E. (1990). *From Etymology to Pragmatics*. Cambridge: Cambridge University Press.

Talmy, L. (1988). Force dynamics in language and cognition. *Cognitive Science*, 12: 49–100.

Tancredi, C. (2007). A multi-model modal theory of I-semantics. part i: Modals. MS., Keio University. <http://semanticsarchive.net/Archive/WFkYjU1Y/>.

Tatevosov, S. (2001). From resultatives to evidentials: Multiple uses of the perfect in Nakh-Daghestanian languages. *Journal of Pragmatics*, 33: 443–64.

Tatevosov, S. (2003). Inferred evidence: Language-specific properties and universal constraints. In Jaszczolt, K. and Turner, K. (eds), *Meaning Though Language Contrast*, vol. 1, pp. 177–92. Philadelphia: Benjamins.

Thomason, R. H. (2002). Combinations of tense and modality. In Gabbay, D. M. (ed.), *Handbook of Philosophical Logic*, vol. 7, 2nd edn. pp. 205–34, Dordrecht: Kluwer Academic Publishers.

Tonhauser, J. (2006). *The Temporal Interpretation of Noun Phrases: Evidence from Guaraní*. PhD thesis, Stanford University.

Traugott, E. C. (2006). Historical aspects of modality. In Frawley, W. (ed.), *The Expression of Modality*, pp. 107–39. Berlin: Mouton de Gruyter.

Traugott, E. C. (1989). On the rise of epistemic meanings in English: an example of subjectification in semantic change. *Language*, 65: 31–55.

Traugott, E. C. and Dasher, R. B. (2002). *Regularity in Semantic Change*. New York: Cambridge University Press.

Traugott, E. C. and König, E. (1991). The semantics-pragmatics of grammaticalization revisited. In Heine, B. and Traugott, E. C. (eds), *Grammaticalization*, pp. 189–218. Amsterdam: Benjamins.

Veltman, F. (1986). Data semantics and the pragmatics of indicative conditionals. In Traugott, E., ter Meulen, A., and Snitzer, A. (eds), *On Conditionals*, pp. 147–68. Cambridge: Cambridge University Press.

Veltman, F. (1996). Defaults in update semantics. *Journal of Philosophical Logic*, 25: 221–61.

Veltman, F. (2005). Making counterfactual assumptions. *Journal of Semantics*, 22: 159–80.

Verhagen, A. (2005). *Constructions of Intersubjectivity. Discourse, Syntax, and Cognition*. Oxford: Oxford University Press.

Verhagen, A. (2006). On subjectivity and "long distance wh-movement". In Athanasiadou, A., Canakis, C., and Cornillie, B. (eds), *Subjectification: Various Paths to Subjectivity*, pp. 323–46. Berlin: Mouton de Gruyter.

Villalta, E. (2000). Spanish subjunctive clauses require ordered alternatives. In Jackson, B. and Matthews, T. (eds), *The Proceedings of Semantics and Linguistic Theory X*. Ithaca, NY: CLC Publications, pp. 239–56.

Warmbrod, K. (1982). A defense of the limit assumption. *Philosophical Studies*, 42: 53–66.

Warner, A. R. (1993). *English Auxiliaries: Structure and History*. Cambridge: Cambridge University Press.

Wekker, H. (1976). *The Expression of Future Time in Contemporary British English*. Amsterdam, New York, and Oxford: North-Holland.

Werner, T. (2006). Future and non-future modal sentences. *Natural Language Semantics*, 14(3): 235–55.

Westmoreland, R. (1995). Epistemic must as evidential. In Dekker, P. M. S. (ed.), *Proceedings of the Tenth Amsterdam Colloquium*, pp. 683–702. Amsterdam: ILLC, University of Amsterdam.

Westmoreland, R. (1998). *Information and Intonation in Natural Language Modality*. PhD thesis, Bloomington, IN: Indiana University.

Willett, T. (1988). A cross-linguistic survey of the grammaticalization of evidentiality. *Studies in Language*, 12(1): 51–97.

Winter, S. and Gärdenfors, P. (1995). Linguistic modality as expressions of social power. *Nordic Journal of Linguistics*, 18(2): 137–66.

von Wright, G. H. (1957). *Logical Studies*. London: Routledge and Kegan Paul.

Wurmbrand, S. (1999). Modal verbs must be raising verbs. In Bird, S., Carnie, A., Haugen, J., and Norquest, P. (eds), *The Proceedings of WCCFL 18*, pp. 599–612. Somerville, MA: Cascadilla.

Yalcin, S. (2007a). Epistemic modals. *Mind*, 116(464): 983–1026.

Yalcin, S. (2007b). Nonfactualism about epistemic modality. MS., Cambridge, MA: MIT.

Yamada, T. (2007a). Logical dynamics of commands and obligations. In Washio, T., Sato, K., Takeda, H., and Inokuchi, A. (eds), *New Frontiers in Artificial Intelligence, JSAI 2006 Conference and Workshops, Tokyo, Japan, June 2006*, vol. 4384 of *Lecture Notes in Artificial Intelligence*, pp. 133–46. Berlin, Heidelberg, New York: Springer.

Yamada, T. (2007b). Logical dynamics of some speech acts that affect obligations and preferences. In van Benthem, J., Ju, S., and Veltman, F. (eds), *A Meeting of Minds: Proceedings of the Workshop on Logic, Rationality and Interaction, Beijing, 2007*, vol. 8 of *Texts in Computer Science*, pp. 275–89. London: King's College Publications.

Zanuttini, R. and Portner, P. (2003). Exclamative clauses: at the syntax-semantics interface. *Language*, 79(1): 39–81.

Zeijlstra, H. (2008). Modal concord is syntactic agreement. In Gibson, M. and Friedman, T. (eds), *The Proceedings of SALT XVII*. Ithaca, NY: CLC Publications, available at <http://ling.auf.net/lingBuzz/000494>.

Zimmermann, T. (2000). Free choice disjunction and epistemic possibility. *Natural Language Semantics*, 8: 255–90.

Zubizarreta, M. L. (1982). *On the Relationship of the Lexicon to Syntax*. PhD thesis, Cambridge, MA: MIT Press.

Index

LIBRARY, UNIVERSITY OF CHESTER